RESOLVING THE
CHOLESTEROL CONTROVERSY

The Scientists Who Proved the Lipid Hypothesis of
Causation of Atherosclerosis and Coronary Heart Disease

RESOLVING THE CHOLESTEROL CONTROVERSY

The Scientists Who Proved the Lipid Hypothesis of Causation of Atherosclerosis and Coronary Heart Disease

GILBERT R. THOMPSON

Imperial College London, UK

W[■] World Scientific

NEW JERSEY · LONDON · SINGAPORE · BEIJING · SHANGHAI · HONG KONG · TAIPEI · CHENNAI · TOKYO

Published by

World Scientific Publishing Europe Ltd.

57 Shelton Street, Covent Garden, London WC2H 9HE

Head office: 5 Toh Tuck Link, Singapore 596224

USA office: 27 Warren Street, Suite 401-402, Hackensack, NJ 07601

Library of Congress Cataloging-in-Publication Data

Names: Thompson, G. R. (Gilbert R.), author.

Title: Resolving the cholesterol controversy : the scientists who proved the lipid hypothesis of causation of atherosclerosis and coronary heart disease / Gilbert R. Thompson.

Description: New Jersey : World Scientific, [2023] | Includes bibliographical references and index.

Identifiers: LCCN 2022056896 | ISBN 9781800613973 (hardcover) |
    ISBN 9781800613980 (ebook for institutions) | ISBN 9781800613997 (ebook for individuals)

Subjects: MESH: Cholesterol, LDL--history | Cholesterol, LDL--adverse effects |
    Atherosclerosis--history | Coronary Disease--history |
    Hypercholesterolemia | Hypolipidemic Agents

Classification: LCC QP752.C5 | NLM QU 11.1 | DDC 612.1/2--dc23/eng/20230317

LC record available at https://lccn.loc.gov/2022056896

British Library Cataloguing-in-Publication Data

A catalogue record for this book is available from the British Library.

For any available supplementary material, please visit
https://www.worldscientific.com/worldscibooks/10.1142/Q0411#t=suppl

Desk Editors: Nimal Koliyat/Adam Binnie/Shi Ying Koe

Typeset by Stallion Press
Email: enquiries@stallionpress.com

Foreword

This is a comprehensive account of the pursuit of the "Lipid Hypothesis," beginning with the pioneering work of Anitschkow and continuing to the present. The narrative is built around individuals who contributed in a positive or negative way to the testing of this hypothesis. In its simplest form the hypothesis is that cholesterol deposition in the arterial wall is a central and integral part of atherosclerosis and its clinical sequelae, heart attacks and strokes. The cholesterol that is deposited is carried in apoB containing lipoproteins, mainly as LDL.

The Swedish physician Carl Muller first described the association of elevated blood cholesterol levels with deposits on the skin called xanthomas and coronary heart disease. John Gofman subsequently made profound contributions in describing the LDL and VLDL fractions as being the lipoproteins associated with increased coronary risk. His experiments were based on the use of the analytical ultracentrifuge to separate lipoproteins.

The National Institutes of Health (NIH)-sponsored Framingham Heart Study identified cholesterol and then LDL-cholesterol (LDL-C) as major risk factors. At the instigation of Donald Fredrickson and Robert Levy, the National Heart and Lung Institute of the NIH launched a series of US lipid research clinics and a coronary primary prevention trial. This trial was the first to show that reducing LDL-C with a drug in hypercholesterolemic men resulted in a decrease in coronary heart disease events.

At about the same time Michael Brown and Joseph Goldstein announced the discovery of the LDL receptor, which regulated LDL clearance from the blood and led to the discovery and development of drugs which reduced LDL-C via an increase in LDL receptor activity. First and foremost were the statin drugs discovered by Akira Endo in Japan.

The Scandinavian Simvastatin Survival Study or 4S showed a decrease in both coronary events and mortality, the latter being the "Holy Grail" cardiologists were pursuing. Just over 10 years later, a meta-analysis by the Cholesterol Treatment Trialists showed that a 1 mmol/l reduction (39 mg/dl) in LDL-C was associated with a 20–22% reduction in cardiovascular events as well as a decrease in total mortality. These results have been validated in a Cochrane Review. Today, with another class of inhibitors working through the LDL receptor, the PCSK9 inhibitors, it is possible to dial down the LDL-C to extremely low levels. Long-term follow-up is needed, but in the short-term there was no observed toxicity. LDL and apoB are now firmly established as having a causal relationship with cardiovascular disease.

Gilbert Thompson has described all of these events, featuring individuals involved in the pursuit of the lipid hypothesis on both sides of the controversy.

Gil enhances the story with his personal observations and contributions. He spent time in my department in Houston in the early 1970s. After performing experiments involving antibodies produced in a goat, he presented me with the Goat's head, which resides in my office, and has the appellation, "Goatto."

Dr. Thompson has pulled together many individuals into a compelling story of the "lipid hypothesis." This book will be useful to anyone interested in the history of cholesterol and cardiovascular disease. It should be required reading for students, fellows and researchers pursuing the field of lipidology. I highly recommend it.

Antonio M. Gotto, M.D., D.Phil
Dean Emeritus of Weill Cornell Medicine,
New York, NY

Preface

"No way of thinking or doing, however ancient, can be trusted without proof."

— Henry Thoreau, *Walden*

The quotation from Thoreau is especially apt to one of the most important and vigorously contested scientific controversies of our time, the lipid hypothesis of atherosclerosis. When proof of the hypothesis eventually came, it transformed the practice of Medicine at every level, from the way physicians monitor and treat their patients with raised cholesterol levels, to the health policies of governments worldwide in the prevention of cardiovascular disease.

I am one of a fast-dwindling generation who knew or met in their lifetimes most of the scientists, many of them now deceased, whose research helped prove or attempted to disprove the lipid hypothesis, and it is appropriate to recall their lives and scientific exploits while I still can. Apart from longevity, my other qualification for this task is that I was involved in lipid research on both sides of the Atlantic during the 1960s, 1970s, 1980s and 1990s, in Boston, Houston, Montreal and here in London, respectively. This was at the time when lipidology was rapidly emerging as an important sub-speciality of mainstream Medicine, especially in the USA, heralding a period of discovery that culminated in the award of the Nobel Prize to Goldstein and Brown in 1985.

The narrative covers a period of just under 100 years, starting with Anitschkow's experiments with cholesterol-fed rabbits in 1913, and it recounts the endeavours and achievements of the leading actors in this protracted scientific drama. The cast includes practitioners of an extraordinarily wide variety of scientific disciplines: pathology, biophysics, epidemiology, nutrition, cardiology, lipidology, genetics, microbiology, pharmacology, clinical trial design and statistics. Most of them believed that cholesterol played a causal role in atherosclerosis and cardiovascular disease, but some dissented strongly from this conclusion.

The breadth of scientific disciplines involved in these studies is matched by the geographical spread of the participants. Anitschkow worked in Russia, Endo discovered the first statin in Japan, their commercial development by Merck took place in the USA and evidence of benefit from lowering cholesterol came from trials conducted in Scandinavia and the UK. The subsequent meta-analysis of these statin trials in 2005 proved the validity of the lipid hypothesis beyond any doubt, and the history of how this all came about is recalled here for posterity.

About the Author

Gilbert R. Thompson (MD, DSc, FRCP) is currently Emeritus Professor of Clinical Lipidology at the Hammersmith Hospital campus of Imperial College London. He is a Past Chairman of both the British Hyperlipidaemia Association (now HEART UK) and the British Atherosclerosis Society. His other achievements include receiving the Lucian Award for Research in Cardiovascular Diseases of McGill University, Montreal, Canada, in 1981, where he was Visiting Professor at the Royal Victoria Hospital, and he was also Visiting Professor at the Mayo Clinic, Rochester NY, USA, in 1997. He was the Founding Editor of *Current Opinion in Lipidology* in 1990 and Associate Editor of the *Journal of Lipid Research* from 1994–1998. He was editor of *Noble Prizes that changed Medicine* (2012) and *Pioneers of Medicine without a Noble Prize* (2014) and the author of *The Cholesterol Controversy* (2008) and of *Medicine My Vocation, Fishing My Recreation* (2020). His main hobby now is fly-fishing, but in 1994 he was the top fund-raising runner for the British Heart Foundation in the London Marathon.

Acknowledgements

My thanks are due to Tony Gotto for his authoritative Foreword and for bringing back happy memories of my time in Houston; to Liz Manson for kindly formatting yet another of my manuscripts; and above all to Caroline Davidson for her skilful and constructive editing, which has not only greatly improved the book but broadened its appeal to include scientists whose speciality is not lipidology as well as non-scientists.

Contents

Chapter 1

Introduction to the Lipid Hypothesis: The Role of Cholesterol in Atherosclerosis

1.1. Brief History of the Lipid Hypothesis

The essence of the Lipid Hypothesis was expressed by Anitschkow in his dictum "Without cholesterol there is no atherosclerosis." The lives and works of the scientists who researched this proposition during most of the 20th and the start of the 21st centuries are described in this book, starting with Anitschkow and Chalatov's experiments with cholesterol-fed rabbits in 1913 and culminating in the Cholesterol Treatment Trialists' Collaborators' meta-analysis of the statin trials in 2005.

In an article published in 2013, Steinberg celebrated the 100th anniversary of the birth of the Lipid Hypothesis, namely that a raised level of cholesterol in the blood causes atherosclerosis.[1] However, the first time the term Lipid Hypothesis was used in the literature was by Ahrens at the Rockefeller University in 1976,[2] who interpreted it as meaning that lowering plasma lipids in patients with raised levels will lead to a reduction in coronary heart disease. Despite his doubts about treatment ("whether rather than how") Ahrens's definition is the corollary of Steinberg's interpretation of Anitschkow's findings, the proof of the pudding being in the eating thereof.

This introductory chapter is aimed at setting the scene for the general reader rather than informing those involved in the specialised field of lipidology. For both sets of readers, a historical account of the scientific research relating to the lipid hypothesis forms the narrative in subsequent chapters.

1.2. Lipids and Lipoproteins

Lipids are not everyone's bread and butter and this section is intended for readers who are not scientists and for scientists whose speciality is not lipidology.

Lipids are fatty substances, the main ones present in our bodies being cholesterol, triglycerides and phospholipids. Linked to apolipoproteins (apo) these lipids are transported in the blood in the form of lipoproteins. The latter are spherical particles with diameters ranging from 6.5 to 1,000 nanometres (nm) (1 nm = one millionth of 1 mm). Their density ranges from <1 to 1.2 grams (g) per millilitre (ml) while the density of plasma in which they circulate is 1.006 g/ml.[3]

Structurally, lipoproteins consist of a complex of lipids which are rendered miscible with the aqueous plasma by their chemical-linkage with various apoproteins, the major ones being apoA-1, apoB and apoE. In addition to performing this function these apoproteins are recognised by and bind to specific receptors in the body that regulate lipoprotein metabolism.

The lipoproteins most relevant to this book are very low density lipo-protein (VLDL), triglyceride-rich particles secreted by the liver, and low density lipoprotein (LDL), the smaller and denser cholesterol-rich product of VLDL. VLDL contains both apoB and apoE, but LDL contains apoB alone and is the main carrier of cholesterol in the bloodstream. The level in the blood of LDL cholesterol (mainly present as cholesterol ester, i.e. linked to a fatty acid) is regulated by the activity of apoB receptors on the surface of cells, which were discovered by Goldstein and Brown in 1974[4] and led to their being awarded the Nobel Prize.

A third class of lipoprotein in blood is high density lipoprotein (HDL), which contains apoA-1 and is the smallest and most dense of the plasma lipoproteins. HDL is rich in phospholipids and appears to exert a protective effect against atherosclerosis. This probably reflects the ability of HDL to act as an acceptor of extracellular cholesterol, e.g. cholesterol located in the arterial wall, which it then transports to the liver for removal by an HDL receptor.

Cholesterol is a Jekyll and Hyde molecule, essential for the healthy function of cells throughout the body and for the synthesis of hormones and bile acids. However, when present in the blood in excess, due to dietary or other life-style influences or genetic disorders that increase levels of VLDL or LDL, cholesterol provokes a pathological process termed atherogenesis. The latter is initiated by the accumulation of cholesterol in the arterial wall, which triggers an inflammatory reaction and leads to the formation of yellowish fatty streaks, the first visible stage of atherosclerosis. Low levels of HDL compound the mischief caused by raised levels of LDL, reflecting the Yin–Yang relationship of these lipoproteins with atherosclerosis.

1.3. Atherosclerosis and Its Clinical Consequences

Atherosclerosis is a disorder that causes narrowing of the lumen of large and medium-sized arteries in various parts of the body, most notably the coronary arteries, which results in coronary heart disease (CHD). The first description of coronary atherosclerosis as a cause of death was in a man with angina by the Scottish surgeon John Hunter in 1776. Two centuries later, CHD was the commonest cause of death in men and women in England and Wales, with more than a quarter of the deaths in men occurring before the age of 65.[5]

A study published in 1972 described a group of patients in London followed up after sustaining a myocardial infarct or heart attack, the chief manifestation of CHD.[6] Almost 50% died within 3 months, half of whom were already dead on arrival at hospital. The mortality rate was greatest in those with the highest plasma cholesterol.

Between 1970–1980, England and Wales had the third highest cholesterol levels and death rates from CHD in the world, exceeded only by Ireland and Finland.[5] The tragic socio-economic consequences of CHD all too common in that era were strikingly captured in a heart charity's fundraising poster of the time (Fig. 1.1).

1.4. The Lipid Hypothesis of Atherosclerosis

The lipid hypothesis proposes that deposition of cholesterol in the inner lining (intima and media) of arteries such as coronary arteries, which supply blood to the heart muscle (myocardium), results in the formation of fatty streaks which then progress to become atheromatous plaques that narrow arteries. Lipid-rich plaques can rupture and cause a clot (thrombus) that blocks a coronary artery, resulting in a heart attack (myocardial infarct). The main risk factor promoting atherogenesis is a raised level of LDL cholesterol in the blood and the validity of the lipid hypothesis

Fig. 1.1.   Poster illustrating the tragedy of premature death from coronary heart disease in middle-aged men in the late 20th century.

depends on demonstrating that the ensuing atherosclerosis can be prevented or reversed by treatment that lowers LDL cholesterol.

Proof that regression of atherosclerosis can occur, aptly described as "atheroexodus" by David Kritchevsky, was confirmed by the discovery of the statins, a novel class of drugs that inhibit cholesterol synthesis and upregulate the activity of LDL receptors in the liver. This increases receptor-mediated removal of LDL from the bloodstream and lowers plasma cholesterol by up to 50%.

Statins were shown to significantly reduce both LDL cholesterol and cardiovascular events (heart attacks and strokes) in a series of clinical outcome trials conducted at the end of the 20th century. The results of those trials hugely influenced public health strategies for managing and preventing coronary heart disease and has resulted in a marked reduction in morbidity and mortality associated with the latter disorder in the USA, UK, Europe and, more recently, other parts of the world.

Recent British Heart Foundation statistics show that prescriptions for lipid-lowering drugs in England, predominantly statins, increased 70-fold between 1991 and 2016,[7] and it is estimated that 7–8 million adults in the UK are now taking a statin. Over the course of those 25 years the UK death rate from coronary heart disease in men and women of all ages decreased by 63%, increased use of statins and revascularisation procedures (coronary artery bypass grafting and angioplasty) both contributing to this beneficial outcome.

In the USA, it was estimated that 39 million adults over the age of 40 were taking a statin in 2012–2013.[8] In another US study, the rate of decrease in serum cholesterol in middle-aged men between 1980–1987 in the pre-statin era was similar to that between 1990–2002, when up to 20% were taking a statin.[9] This suggests firstly that the US diet was exerting a cholesterol-lowering effect prior to the introduction of statins, and secondly that the lack of any accentuation of the rate of decrease in serum cholesterol after the introduction of statins implied that they were being used as a substitute for rather than as an adjunct to diet. This explanation is supported by a subsequent study that showed that intake of calories and fat increased among statin users over time compared with nonusers, which underlines the importance of statin users maintaining dietary restraint.[10]

1.5. The Author's Scientific Background

The purpose of the following summary of my experiences and credentials is to define my perspective of the science and the scientists who tested, contested and eventually proved the lipid hypothesis of atherosclerosis. In other words, to illustrate where I'm coming from.

My first real contact with lipidology came in 1966 when I was awarded a Medical Research Council (MRC) Travelling Fellowship to spend a year as a Research Fellow in Medicine at Harvard Medical School. My supervisor was Kurt Isselbacher and his laboratory was situated at the Massachusetts General Hospital in Boston. My research project was to investigate the effect of different types of dietary fat on the absorption of ^{14}C-cholesterol and ^{3}H-vitamin D3 in rats with a surgically created lymph fistula.[11] The latter procedure, skilfully performed by my American co-workers, prevented intestinal lymph from entering the bloodstream by diverting its main duct to the exterior, enabling the absorption from the small intestine via the lymphatic pathway of lipid-soluble compounds such as cholesterol and vitamin D to be measured directly in lymph collected from the fistula. This involved my acquiring basic biochemical skills such as Folch extraction of lipids from samples of lymph, using chloroform: methanol 2:1 v/v as the solvent, performing thin layer chromatography to separate free from esterified cholesterol and vitamin D and then assaying the radioactivity in lipid extracts in a liquid scintillation counter.

My initial training was in gastroenterology but, as described elsewhere,[12] my interest increasingly turned towards lipidology after I returned from Boston in 1967, mainly because of what seemed to me to be an obvious but largely ignored link between cholesterol and CHD in the UK. I visited the director of the MRC Clinical Research Centre at Northwick Park to enquire whether there would be an opening there to work on lipids only to be told that "cholesterol was not the soft underbelly of research into atherosclerosis."

Instead, I negotiated a year's leave of absence from Hammersmith Hospital (the Hammersmith) and returned to the States in 1972–1973 to work as an Assistant Professor in Tony Gotto's Division of Atherosclerosis and Lipoprotein Research at Baylor College of Medicine in Houston. His research was partly funded by the Lipid Clinics Research Program of the National Heart Lung and Blood Institute (NHLBI) and the laboratory was very well equipped. Most of those working there were biochemistry postgraduates but only a minority were Texans. Houston was booming in the early 70s and academic institutions like Baylor were attracting scientists from all over the USA to undertake research on lipids and atherosclerosis.

While in Houston I visited David Bilheimer in nearby Dallas and learnt how to radio-iodinate apoB with ^{125}I and with [131]I. This enabled me to develop a radioimmunoassay for measuring apoB in plasma, using antibodies raised in goats,[13] and to study the turnover of ^{125}I-apoB in human subjects.[14] During my stay in Houston, which lasted a very rewarding 18 months, I also learnt how to isolate lipoproteins of differing density by preparative ultracentrifugation of plasma in aluminium-capped plastic tubes with sequential spins at 40,000 rpm for 16 hours in a fixed angle rotor, using a tube slicer (a horizontal guillotine) to separate lipoproteins in the supernatant from those in the infranatant at the end of each run.

These basic skills proved invaluable in the research I pursued when I returned to London to join the MRC Lipid Metabolism Unit at the Hammersmith Hospital and again later on when I worked with Allan Sniderman at the Royal Victoria Hospital in Montreal.[15]

Another source of knowledge came from attendance at scientific meetings. Between 1972 and 1998 I participated in a total of more than

120 meetings, roughly one in every 3 months, which were held in cities all over the world, often in the USA. Drug company sponsorship was far less restricted in those days than it is now and I seldom had to pay for my travel and hotel accommodation.

These meetings provided an opportunity to present one's research to one's peers and to discover what they were working on. The best meetings were those organised by the Council on Arteriosclerosis of the American Heart Association (AHA), which took place during the annual AHA meetings in November. They enabled me to assess who was who in the lipid hierarchy, ranking them according to the quality of their presentations and their performance in the discussion that followed.

As time went by, I established my own reputation, which resulted in a steadily increasing number of invitations to speak at lipid meetings. This in turn led to requests to act as a referee for research grant proposals and to review manuscripts submitted for publication in various scientific journals, while at the same time submitting my own research to similar scrutiny. Medical research is a competitive pursuit and any perceived imbalance in the narrative between the plethora of scientific references and paucity of biographical ones reflects the fact that I was often more familiar with the published work of the scientists than with the scientists themselves.

I achieved sufficient recognition in the UK to be elected Chairman of both the British Hyperlipidaemia Association (now HEART UK) and the British Atherosclerosis Society. Other achievements included receiving the Lucian Award for research in cardiovascular diseases of McGill University, Montreal in 1981, where I was a Visiting Professor for 3 months. I became the founding Editor of *Current Opinion in Lipidology* in 1990, an Associate Editor of the *Journal of Lipid Research* in 1994 and I was a member of the British Heart Foundation Project Grants Committee between 1991 and 1994. I was appointed Professor of Clinical Lipidology at the Royal Postgraduate Medical School (now Imperial College London) in 1998 and am now an Emeritus Professor in the Faculty of Medicine at the Hammersmith Hospital campus of Imperial College.

1.6. Criteria for Inclusion of Scientists in the Book

The three most prestigious awards in biomedical science are the Nobel Prize for Physiology or Medicine, the Lasker Awards and the Gairdner Awards. The award of any one of these prizes provides objective evidence that the recipient is acknowledged by the scientific community as being in the top rank of biomedical scientists. Belonging to this category ensures recognition in this history of the lipid hypothesis and includes the recipients of two Nobel Prizes, three Lasker Awards and five Gairdner Awards.

When it came to selecting the remainder of the scientists who feature in the book, subjective judgement played a greater role. The main criterion was whether the scientists concerned had made a major contribution to proving or disproving the lipid hypothesis of atherosclerosis. Over 50% of those selected worked in the USA, reflecting the emphasis there on research into cardiovascular disease from the 1950s onwards.

The research described in the following chapters is in a roughly historical sequence. To put it into its geographical context, the locations where each of the scientists worked and their nationalities are listed in Table 1.1. A glossary of scientific terms used in the narrative is appended at the end of the book.

Table 1.1.   Alphabetical list of scientists, their nationalities and research locations.

Researchers	Nationality	Location
Anitschkow	Russian	St. Petersburg, Russia
Berg	Norwegian	Oslo, Norway
Dawber & Kannel	American	Framingham, Massachusetts, USA
Endo	Japanese	Tokyo, Japan
Fredrickson & Levy	American	Bethesda, Maryland, USA
Gofman & Havel	American	San Francisco, California, USA
Goldstein & Brown	American	Dallas, Texas, USA
Keys	American	Minneapolis, Minnesota, USA
McMichael & Mitchell	British	London; Oxford & Nottingham, UK
Myant	British	London, UK
Oliver	British	Edinburgh, UK
Pedersen, Shepherd & Collins	Norwegian, British (2)	Oslo, Norway; Glasgow, UK; Oxford, UK
Steinberg	American	La Jolla, California, USA
Utermann	German	Innsbruck, Austria
Vagelos, Alberts & Tobert	American (2), British	Rahway, New Jersey, USA
Yudkin	British	London, UK

References

1. Steinberg D. In celebration of the 100th anniversary of the lipid hypothesis of atherosclerosis. *J. Lipid Res.* 2013; **54**: 2946–2949.
2. Ahrens EH Jr. The management of hyperlipidemia: Whether, rather than how. *Ann. Intern. Med.* 1976; **85**: 87–93.
3. Thompson GR. *A Handbook of Hyperlipidaemia.* London, UK: Current Science; 1994.
4. Goldstein JL, Brown MS. Binding and degradation of low density lipoproteins by cultured human fibroblasts. *J. Biol. Chem.* 1974; **249**: 5153–5162.
5. Thompson GR, Wilson PW. *Coronary Risk Factors and Their Assessment.* London, UK: Science Press; 1992.
6. Patterson D, Slack J. Lipid abnormalities in male and female survivors of myocardial infarction and their first-degree relatives. *Lancet* 1972; **1**(7747): 393–399.
7. British Heart Foundation Heart and Circulatory Diseases Statistics 2018. www.bhf.org.uk.
8. Salami JA, Warraich H, Valero-Elizondo J, *et al.* National trends in statin use and expenditures in the US adult population from 2002 to 2013. *JAMA Cardiol.* 2017; **2**: 56–65.
9. Arnett DK, Jacobs DR, Luepker RV, *et al.* Twenty-year trends in serum cholesterol, hypercholesterolemia, and cholesterol medication use. *Circulation.* 2005; **112**: 3884–3891.
10. Sugiyama T, Tsugawa Y, Tseng C-H, Kobayashi Y, Shapiro MF. Different time trends of caloric and fat intake between statin users and nonusers among US adults: Gluttony in the time of statins? *JAMA Intern. Med.* 2014; **174**: 1038–1045.
11. Thompson GR, Ockner RK, Isselbacher KJ. Effect of mixed micellar lipid on the absorption of cholesterol and vitamin D_3 into lymph. *J. Clin. Invest.* 1969; **48**: 87–95.
12. Thompson G. *Medicine My Vocation, Fishing My Recreation.* London, UK: World Scientific Publishing Europe; 2020.
13. Thompson GR, Birnbaumer ME, Levy RI, Gotto AM Jr. Solid phase radioimmunoassay of apolipoprotein B (apo B) in normal human plasma. *Atherosclerosis* 1976; **24**: 107–118.
14. Thompson GR, Jadhav A, Nava M, Gotto AM Jr. Effects of intravenous phospholipid on low density lipoprotein turnover in man. *Eur. J. Clin. Invest.* 1976; **6**: 241–248.
15. Teng B, Sniderman AD, Soutar AK, Thompson GR. Metabolic basis of hyperapobetalipoproteinemia. Turnover of apolipoprotein B in low density lipoprotein and its precursors and subfractions compared with normal and familial hypercholesterolemia. *J. Clin. Invest.* 1986; **77**: 663–672.

Chapter 2

Nikolai Anitschkow: The Cholesterol-Fed Rabbit, an Animal Model of Atherosclerosis*

2.1. Introduction

Timing, they say, is all important. If a discovery is too far ahead of itself, if it is not fully appreciated by one's contemporaries, it will probably go unrewarded. That was the case with Nikolai Nikolaevich Anitschkow's 1913 formulation of the hypothesis that deposition of cholesterol in the artery wall is the initiating factor in atherogenesis and coronary heart disease — the so-called lipid hypothesis.[1,2] What he showed, with his undergraduate medical student collaborator, Semen Chalatov, was that simply feeding rabbits pure cholesterol dissolved in vegetable oil, and thus raising their blood cholesterol concentrations to very high levels, was sufficient to produce arterial lesions very similar to those of human atherosclerosis. No other intervention was needed — no infection, no induced inflammation, no injury, no change in blood pressure, no renal dysfunction, no diabetes — just a rise in blood cholesterol level.[3]

Over the next two decades, a series of classic studies in Anitschkow's laboratory characterised the rabbit model in remarkable detail.

---

*Edited version of chapter by Steinberg D. Anitschkow: Birth of the lipid hypothesis of atherosclerosis and coronary heart disease. In: *Pioneers of Medicine without a Nobel Prize*, Thompson G (ed.). London: Imperial College Press; 2014.

His summary of that work, published in 1933,[4] proposed the use of his rabbit model to explore the pathogenesis of atherosclerosis, predicting that from such studies would come insights into the nature of the human disease. Sad to say, his findings were ignored and four decades passed before the importance of Anitschkow's work was fully realised and serious study of atherogenesis began. Why was his animal model rejected as irrelevant to human disease? And why was there scepticism regarding the role of hypercholesterolaemia in atherosclerosis?

2.2. How Significant Were His Findings?

Anitschkow's cholesterol-fed rabbits represented the first animal model for one of the major causes of human morbidity and mortality — coronary artery disease due to atherosclerosis. The blood cholesterol levels of his rabbits rose almost immediately and within weeks their arteries began to show raised yellow lesions rich in "lipoids".

According to Anitschkow: "The blood of such animals exhibits an enormous increase in cholesterin [cholesterol] content, which in some cases amounts to several times the normal quantity. It may therefore be regarded as certain that in these experimental animals large quantities of the ingested cholesterin are absorbed, and that the accumulations of this substance in the tissues can only be interpreted as deposits of lipoids circulating in large quantities in the humours of the body."

Although it was at the time largely ignored and considered irrelevant to the human disease, Anitschkow's model ultimately became the foundation for much of modern atherosclerosis research. A major reason for the scepticism with which his work was met was that many of the attempts to repeat his work were made in either rats or in dogs. These were the animals most commonly in use in laboratories of experimental physiology at the time. Feeding cholesterol to these animals did not result in the development of arterial disease. What was not appreciated was that feeding cholesterol to these animal species does not raise their blood cholesterol levels.

Anitschkow himself was aware that dogs and rats were for some reason resistant to high cholesterol intake in the diet. Presciently, he proposed that, being carnivores, they had a large capacity to handle cholesterol in the diet, converting it rapidly to bile acids for excretion. However, he also knew that rabbits were not a uniquely susceptible species. Guinea pigs and goats responded like rabbits and developed both hypercholesterol-aemia and atherosclerosis.[5] Later it was shown that dogs would develop atherosclerosis on a high cholesterol diet *if* you first rendered their thyroid gland underactive. The resulting hypothyroidism caused down regulation of the LDL receptors so that the high cholesterol diet now *did* raise blood cholesterol sufficiently and now *did* induce atherosclerosis.[6]

2.3. State-of-the-Science in 1913

By 1913 it was well recognized by pathologists that degenerative changes in the arteries (atherosclerosis or arteriosclerosis, terms now used synony-mously) was one of the hallmarks of ageing. However, the pathogenesis of the disease was not understood. Most investigators favoured the view that the lesions must be due to some form of "injury," but the nature of that "injury" remained undefined.

In his 1933 review, Anitschkow lists more than a dozen papers describing early efforts to induce atherosclerotic lesions using mechanical or chemical injury to the artery wall (ligation, pulling, pinching, wound-ing, cauterization with silver nitrate or galvanic wire).[4] He sums up the results as follows: "… these different ways of causing lesions in the arter-ies resulted merely in the production of inflammatory arterial changes which had no similarity with human arteriosclerosis." He went on to note that such interventions did alter the wall of the artery in such a way as to make it more susceptible to lesion formation if there was concomitant hypercholesterolaemia. He cites the work of Ssolowjew who showed that injury such as cauterization could lead to "regenerative thickening of the intima" (inner lining of arteries) but no lipid deposition. If, however, cho-lesterol feeding was started before the injury was produced, the site of the injury showed more severe atherosclerotic changes and greater lipid deposition.[7]

Anitschkow's conclusion was that "… lesions of the arterial wall which are associated with hyperplasia of the intima create a local predis-position to the formation of lipoidal deposits, that is, to the development of atherosclerosis, provided that they are synchronous with hypercholes-terolemia …." This prescient conclusion has been borne out by many later studies differentiating lesions due to arterial injury created without raising cholesterol from those produced in the presence of increased levels.

2.4. Anitschkow's Observations

Anitschkow was a keen observer and an insightful experimenter. Having established that simply feeding pure cholesterol to rabbits was sufficient to induce human-like arterial lesions (Fig. 2.1), he proceeded to explore the natural history of those lesions and the factors that influenced their development.

Over the next two decades he and his colleague Chalatov established many of the key elements in the pathogenesis of atherosclerosis.[1–3] They showed that:

(1) In the earliest lesions, the fatty streaks, most of the lipid was found in cells containing large numbers of vacuoles. Because lipids are extracted during the routine preparation of tissue samples, the multiple lipid droplets are seen as empty vacuoles; hence the designation "foam cells."

(2) In frozen sections the lipid droplets were birefringent (refracted light under the microscope). Anitschkow recognised this as representing liquid crystals of cholesterol esters.

(3) The cholesterol ester-loaded foam cells were probably white blood cells that had infiltrated the artery wall. Thus, he anticipated that inflammation might play a role in lesion development.

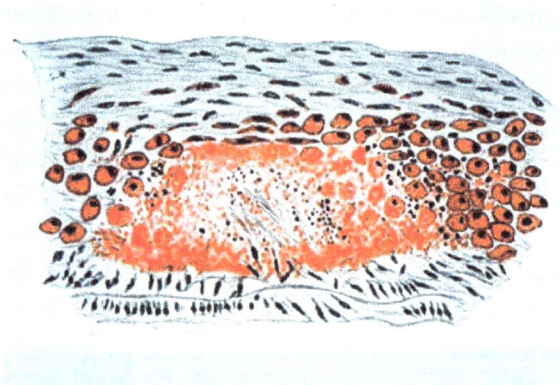

Fig. 2.1.   A highly magnified advanced lesion in a cholesterol-fed rabbit drawn by Anitschkow. He noted the central necrotic core containing needle-like crystals of cholesterol, the foam cells surrounding the core and the fibrous cap overlying the lesion.[4]

(4) The monolayer of endothelial cells over the lesions appeared to be intact, indicating that the invading blood cells must penetrate between the endothelial cells. Thus, endothelial denudation, while it clearly did occur at a later time, was not a necessary antecedent to lesion formation.

(5) There was a characteristic pattern of distribution of early lesions. They were most severe at arterial branch points and Anitschkow correctly surmised that this localisation was most likely determined by haemodynamic factors. In these rabbits the very earliest lesions appeared at the root of the aorta and in the aortic arch and then proceeded distally, being located above the orifices of arterial branches.

(6) Over long periods of cholesterol feeding there was ultimately deposition of connective tissue and development of a fibrous cap over the lipid core of the atheroma (conversion of a fatty streak to a fibrous plaque).

(7) Early lesions were partially reversible but the reversal was a very slow process. Most, but not all, of the lipids could be mobilized from advanced lesions, leaving behind the fibrous cap and a few cholesterol crystals.

(8) The extent of lesions was proportional to the degree of blood cholesterol elevation and the duration of exposure to it. Anitschkow was well aware that it was the level of *blood* cholesterol that determined the size and extent of lesions, not necessarily the amount of cholesterol ingested.

(9) While the blood cholesterol level is critically important, other factors can and do play a significant part. Anitschkow's dictum "No atherosclerosis without cholesterol" has often been cited as evidence that he was unaware of the multifactorial nature of the disease. However, his 1933 review disproves this. There he sums up as follows: "The views here set forth concerning the etiology of atherosclerosis constitute what I have called the 'combination theory' of its origin." So, it should be clear that he was fully aware that the degree of atherosclerosis, while perhaps most evidently dependent on the degree of blood cholesterol elevation, could be significantly affected by other factors such as blood pressure, toxic substances, local arterial changes and the like.

In his rabbit model, however, no such additional insults or injuries were needed; hypercholesterolaemia was a *sufficient* cause. The correctness of this conclusion was most dramatically underscored by Yoshio Watanabe's discovery in 1980 of a strain of rabbits that have blood cholesterol levels around 600 mg/dl and uniformly develop atherosclerosis *on a normal rabbit diet*.[8] These rabbits have a mutation of the LDL receptor gene identical to that found in some patients with familial hypercholesterolaemia.[9] Thus, impaired uptake of LDL by the LDL receptor leads to high LDL cholesterol levels which in turn initiates atherogenesis.

It is quite remarkable how well Anitschkow's description of atherogenesis has stood the test of time. While there have been many advances at the level of biochemistry, cell biology and molecular biology, the basic pathogenesis as described 100 years ago by this experimental physiologist from St. Petersburg requires little or no amendment.

2.5. Biography

Nikolai Nikolaevich Anitschkow was born in 1885 in St. Petersburg, the son of a prominent educator who served as a Vice-Minister of Education in the Russian Empire, while his mother was the daughter of a Russian Orthodox priest. In 1903, Anitschkow entered the Imperial Military Medical Academy in St. Petersburg. There he became a protégé of Alexander Maximow, professor of histology and author of what was at the time one of the most respected textbooks of histology in the world. Anitschkow did his thesis research on myocardial inflammation and in the course of that work described a unique cell type that is still known today as the "Anitschkow myocyte."

He succeeded with his very first project — the demonstration that cholesterol itself could raise cholesterol levels in the blood of rabbits and induce arterial changes very similar to those seen in human atherosclerosis, as discussed above.[1,2] Shortly after graduation he went to Strasbourg to study with Hans Chiari and then to Freiburg to work with Karl A. L. Aschoff. These were the outstanding centres of experimental pathology in Europe. Both Chiari and Aschoff were much impressed by Anitschkow's work and encouraged him to continue and expand his project.

In August 1914, war broke out and Anitschkow was called to active duty with the Russian Army Medical Corps in St. Petersburg. After the war he was appointed head of the department of pathological physiology at his alma mater, the Military Medical Academy in St. Petersburg, a position he held until 1939. He continued to explore the pathogenesis of rabbit atherosclerosis in detail and was invited to present his findings at several international congresses during the 1920s and 1930s. So, it cannot be argued that his colleagues were unaware of his findings. He intended to write a book pulling together his data and his ideas on atherogenesis and was encouraged to do so in a letter from Aschoff written in 1929: "How is your work on your large monograph on atherosclerosis? ... you are truly destined to write a great summary work on atherosclerosis." Unfortunately, Anitschkow never did complete the monograph but his comprehensive 52-page article in Cowdry's compendium *Arteriosclerosis: A Survey of the Problem*, published in 1933, tells the story well.[4]

Anitschkow was highly respected not only as a scientist but also as a capable administrator. He headed the two departments of pathology at which he worked in St. Petersburg (renamed Leningrad in 1924) and was elected President of the Academy of Medical Sciences of the USSR (1946–1953). He died on December 7, 1964. The cause of death, ironically, was myocardial infarction.

2.6. Ancillary Research

In 1962, just 2 years before he died, Anitschkow established a new Laboratory of Lipid Metabolism in his Institute of Experimental Medicine and picked Dr. Anatoly N. Klimov to head it. Klimov had previously worked on lipoproteins at the US National Institutes of Health as a World Health Organization fellow. In 1966, Klimov and his colleagues reported a heroic experiment in which they purified lipoprotein-rich serum from cholesterol-fed rabbits and transfused it into rabbits fed only a normal diet. Because serum lipoproteins have a relatively short half-life, it was necessary to process and infuse huge amounts of donor lipoproteins to keep the level of serum cholesterol sufficiently high in the recipients.[10] Each rabbit received 14 to 25 grams of lipoprotein cholesterol intravenously over a period of 5 to 7 months. They developed significant arterial lesions, nicely confirming Anitschkow's hypothesis that the direct cause of atherosclerotic lesions in the cholesterol feeding experiments is simply — hypercholesterolaemia. There is something satisfying about the continuity of research in St. Petersburg, from the first rabbit feeding experiments of 1913 to the direct demonstration of causality in 1966.

2.7. The Impact of Anitschkow's Research

As discussed above, his work certainly did not make a great splash in the early 20th century. Quite the opposite. At the time most of his colleagues thought it was an odd result, almost certainly irrelevant to human disease. On the other hand, his findings were soon confirmed by a few laboratories using rabbits[11] and guinea pigs[12] and gradually the cholesterol-fed rabbit became a widely used model.

Anitschkow's findings were the inspiration for John Gofman and his pioneering studies on human lipoprotein fractions and their correlation with coronary artery disease[13,14] as discussed in Chapter 3. Anitschkow's findings also figured large in the thinking of Ancel Keys[15] and of Kannel, Dawber and the rest of the group that initiated the classic Framingham studies,[16] studies that suggested that the chances of having a heart attack increased with increases in blood cholesterol concentration.

In 1958, William Dock, an eminent cardiologist, reviewed "the first fifty years" of atherosclerosis research.[17] He likened Anitschkow's discovery to the discovery of the tubercle bacillus by Koch. In their 1998 book, *Medicine's 10 Greatest Discoveries*,[18] Friedman and Friedland include Anitschkow's discovery of the role of cholesterol in atherogenesis alongside Fleming's discovery of penicillin and Jenner's discovery of vaccination. They make clear their admiration for his scientific contributions but include some comments on his politics. They say that "… he remained a loyal, indeed doctrinaire, member of the Communist Party and a friend of Joseph Stalin." Referring to a photograph of Anitschkow taken in 1945 (Fig. 2.2), they comment "One need only peruse his Slavic face, with its high forehead and cheekbones, to recognize it as the visage of a man who was neither gregarious nor benevolent. Indeed, his was the face of a full-fledged Communist of 1918 Bolshevist provenance." Why two American physicians who lived through the McCarthy Era felt the issue was relevant to Anitschkow's scientific contributions is not entirely clear.

The unsubstantiated statement regarding Anitschkow's Party membership is flatly contradicted in a 2006 article by Igor E. Konstantinov, Nicolai Mejevoi and Nikolai M. Anichkov.[19] The last co-author is a grandson of Nikolai N. Anitschkow. Noting that "It is difficult to draw conclusions about someone's personality almost a half-century after his death [from a photograph]," they go on to aver that … "Anitschkow was a descendant of Russian nobility … and never joined the Communist Party."

Fig. 2.2.   Nikolai Nikolaevich Anitschkow (1885–1964).
Source: Khavkin and Pozharski.[24]

2.8.   What Led Anitschkow to Feed Rabbits Cholesterol?

There is a very interesting background story to tell here. Like so many breakthroughs in science, Anitschkow's discovery was serendipitous. Many have probably assumed, incorrectly, that Anitschkow was led to do his experiment by the Nobel Prize winning German chemist Adolf Windaus's 1910 paper reporting that the aortas of patients with atherosclerosis contained much higher concentrations of cholesterol than did normal aortas.[20] Actually, the rationale for Anitschkow's studies, as pointed out by Hoeg and Klimov,[21] came from a 1909 paper by Alexander Ignatowski, who was an Associate Professor of Internal Medicine at the Military Medical Academy in St. Petersburg.[22] Ignatowski was following up on a proposition put forward some years earlier by Nobel Prize-winning microbiologist, Ilya Metschnikow, who had postulated that an excess of protein in the diet was potentially toxic and somehow accelerated the ageing process. Ignatowski decided to feed rabbits a protein-rich diet and look for signs of toxicity or premature ageing. He fed his rabbits large amounts of meat, eggs and milk. These diets were indeed toxic in young rabbits, affecting liver and adrenals, but in adult rabbits the major effect was the development of striking arterial lesions resembling those of human atherosclerosis. Since atherosclerosis was considered one of the hallmarks of ageing, Ignatowski considered that his findings had confirmed Metschnikow's "protein toxicity" hypothesis.

One of Ignatowski's research fellows, N. W. Stuckey, extended these studies and showed that the protein in the diet used by Ignatowski was actually not necessary.[23] Egg yolk alone gave the same results as did the feeding of beef brain. It remained for Anitschkow and Chalatov to further narrow things down.[2] They extracted cholesterol from the egg yolks and showed that feeding this pure cholesterol by itself could reproduce the vascular damage induced by the complicated diets rich in eggs, milk or meat. They showed that protein was not the culprit but a lipid — cholesterol. From Metschnikow's point of view this would have been another instance of an ugly fact destroying a beautiful hypothesis, but from Anitschkow's, it was the serendipitous discovery of a valuable new hypothesis, the lipid hypothesis of atherosclerosis.

2.9. Conclusions

Anitschkow's rabbit experiments did not prove the hypothesis he started from — that high protein diets caused premature ageing. However, he realised that the cholesterol–atherosclerosis connection might be highly relevant to the human disease and he followed the scent. Louis Pasteur observed, "Chance favours the prepared mind." Anitschkow's mind was prepared and today everyone in the field acknowledges him as the grandfather of atherosclerosis research.

References

1. Anitschkow NN, Chalatov S. Ueber experimentelle Choleserinsteatose und ihre Bedeutung fur die Entstehung einiger pathologischer Prozesse. *Zentralbl. Allg. Pathol.* 1913; **24**: 1–9.
2. Anitschkow N. Ueber die Veranderungen der Kaninchenaorta bei experimenteller Cholesterinsteatose. *Beitr. Pathol. Anat.* 1913; **56**: 379–404.
3. Classics in arteriosclerosis research: On experimental cholesterin steatosis and its significance in the origin of some pathological processes by N. Anitschkow and S. Chalatow, translated by Mary Z. Pelias, 1913. *Arteriosclerosis* 1983; **3**: 178–182.

4. Anitschkow N. Experimental atherosclerosis in animals. In: *Arteriosclerosis*, Cowdry EV (ed.). New York: Macmillan; 1933.
5. Anitschkow N. Ueber die experimentelle Atherosklerose der Aorta beim Meerschwinchen. *Beitr. Pathol. Anat.* 1922; **70**: 265–281.
6. Steiner A, Kendall FE. Atherosclerosis and arteriosclerosis in dogs following ingestion of cholesterol and thiouracil. *Arch. Path* 1946; **42**: 433–444.
7. Ssolowjew A. Experimentelle Untersuchungen uber die Bedeutung von lokaler Schadigung fur die Lipoidablagerung in der Arterienwand. *Zeitschrift fur die gesamte experimentelle Medizin* 1929; **69**: 94–104.
8. Watanabe Y. Serial inbreeding of rabbits with hereditary hyperlipidemia (WHHL-rabbit). *Atherosclerosis* 1980; **36**: 261–268.
9. Kita T, Brown MS, Watanabe Y, *et al*. Deficiency of low density lipoprotein receptors in liver and adrenal gland of the WHHL rabbit, an animal model of familial hypercholesterolemia. *Proc. Natl. Acad. Sci. USA* 1981; **78**: 2268–2272.
10. Klimov AN, Rodionova LP, Petrova-Maslakova LG. Experimental atherosclerosis induced by repeated intravenous administration of hypercholesterolaemic serum. *Cor. Vasa.* 1966; **8**: 225–230. PMID: 5916583.
11. Bailey CH. Atheroma and other lesions produced in rabbits by cholesterol feeding. *J. Exp. Med.* 1916; **23**: 69–84.
12. Bailey CH. Observations on cholesterol-fed guinea pigs. *Proc. Soc. Exp. Biol. Med.* 1915; **13**: 60–62.
13. Gofman JW, Lindgren F, Elliott H, *et al*. The role of lipids and lipoproteins in atherosclerosis. *Science* 1950; **111**: 166–171.
14. Gofman JW. Serum lipoproteins and the evaluation of atherosclerosis. *Ann. N. Y. Acad. Sci.* 1956; **64**: 590–595.
15. Keys A. Diet and the epidemiology of coronary heart disease. *J. Am. Med. Assoc.* 1957; **164**: 1912–1919.
16. Kannel WB, Dawber TR, Kagan A, *et al*. Factors of risk in the development of coronary heart disease — Six year follow-up experience. The Framingham Study. *Ann. Intern. Med.* 1961; **55**: 33–50.
17. Dock W. Research in arteriosclerosis — the first fifty years. *Ann. Intern. Med.* 1958; **49**: 699–705.
18. Friedman M, Friedman G. *Medicine's 10 Greatest Discoveries*. New Haven: Yale University Press; 1998.
19. Konstantinov IE, Mejevoi N, Anichkov NM, Nikolai N. Anichkov and his theory of atherosclerosis. *Tex. Heart Inst. J.* 2006; **33**: 417–423.
20. Windaus A. Uber den Gehalt nirmaler und atheromatoser Aorten an Cholsterin und Cholesterinestern. *Hoppe-Seyler Z. Physiol. Chemie.* 1910; **67**: 174–176.
21. Hoeg JM, Klimov AN. Cholesterol and atherosclerosis: "The new is the old rediscovered." *Am. J. Cardiol.* 1993; **72**: 1071–1072.

22. Ignatowski A. Uber die Wirkung des Tierischen Eiweisses auf die Aorta und die paerenchymatosen Organe der Kaninchen. *Virchows Arch fur path Anat* 1909; **198**: 248–270.
23. Stuckey NW. Ueber die Veranderungene der Kaninchenaorta bei der Futterung mit verschiedenen Fettsorten. *Centralblatt fur Allgemeine Pathologie und Pathologische Anatomie* 1912; **23**: 910–911.
24. Khavkin TN, Pozharski KM. Nikolai Nikolaevich Anitschkow. *Beitr. Z. Path.* 1975; **150**: 301–312.

Chapter 3

John Gofman and Richard Havel: The Discovery that Cholesterol is Transported by Lipoproteins in the Blood and their Centrifugal Analysis

3.1. Introduction

Dan Steinberg considered that Anitschkow was the grandfather of the lipid hypothesis of atherosclerosis[1] and in a similar manner Virgil Brown dubbed John Gofman the "father of clinical lipidology."[2] This accolade reflects Gofman's discovery that cholesterol is transported in the bloodstream in the form of lipoprotein particles and that concentrations of low density lipoproteins are increased in myocardial infarct survivors; his findings suggested a possible mechanism for the causation of atherosclerosis, both in man and the cholesterol-fed rabbit, namely that the cholesterol in atheromatous plaques is derived from lipoproteins circulating in the plasma.

3.2. Gofman: Biography

John Gofman (1918–2007) was born in Cleveland, Ohio, the son of Jewish immigrants who had fled from Czarist persecution in Russia. He graduated from Oberlin College, Ohio, with a Bachelor of Arts degree in chemistry in 1939 and completed 1 year of medical school before deciding that he needed to learn more chemistry. It was a fateful decision, because he entered the graduate school at the University of California (UC), Berkeley in 1940 at a time when it was the centre for high energy physics research. In 1939, Ernest Lawrence was awarded the Nobel Prize in Physics for the invention of the cyclotron and Gofman's supervisor at Berkeley, Glenn Seaborg, used one to discover Plutonium, a discovery that earned him the Nobel Prize in Chemistry in 1951.

Gofman obtained his PhD in 1943 in nuclear and physical chemistry for the discovery of uranium-232 and uranium-233, and for proving that uranium-233 is fissionable. Because of the sensitive nature of this research, his work became shrouded in the secrecy accompanying the Manhattan Project. This was directed by the UC Berkeley physicist J. Robert Oppenheimer who recognised that Gofman was a talented chemist and asked him to isolate enough Plutonium for urgent experiments at the project's laboratory in Los Alamos, New Mexico.

After completing this task, Gofman resumed studying Medicine and enrolled in what is now UC San Francisco (UCSF) in 1943. He graduated there with an M.D. in 1946.

After serving a one-year internship at UCSF, he joined the Division of Medical Physics in the UC Berkeley Department of Physics. He formed the lipoprotein laboratory in the Donner Laboratory, where he focused on a difficult but important problem — the relationship between blood lipids such as cholesterol and heart disease. Drawing upon the physico-chemical know-how he had acquired while working on the Manhattan Project, he utilised one of the first analytical ultracentrifuges in the United States to isolate the various classes of lipoprotein particles present in blood plasma.

Gofman was recognized at Berkeley as being a dynamic lecturer, superb teacher and generous research mentor and was promoted Professor in the Department of Physics in 1954. In an interview conducted by the US Department of Energy in 1994, set against the background of his research on the biological effects of radiation, Gofman stated: "I didn't have any good ideas about cancer which I thought I might work on. But I had one idea about heart disease and cholesterol, which was poo-pooed at the time as … nonsense. I felt maybe the reason why it … had such a bad name is that people didn't have the technology to study *how* blood transports cholesterol."[3] His decision to study heart disease and cholesterol in 1949 was triggered by Anitschkow's findings in cholesterol-fed rabbits, which intrigued him.

In 1963, Gofman discontinued his research into lipoproteins to set up the medical division at the Lawrence Livermore National Laboratory (LLNL), the US's designer of nuclear weapons, and began investigating the biological effects of radiation. In 1969, he and his colleague at Livermore, Arthur Tamplin, caused controversy when they concluded that the risk of cancer from radiation was far greater than the estimates the US government was using to limit public exposure. He was tireless in pressing the Federal government to drastically revise standards for radiation exposure, to limit nuclear tests and to cancel projects to use nuclear explosives for digging harbours and canals. In 1971, Gofman and Tamplin called for a five-year moratorium on the licensing of new nuclear power plants so that the public health consequences could be studied and urged doctors to reduce unnecessarily high doses of diagnostic X-rays.

Gofman's stand on radiation resulted in the loss of his research funding at the Livermore Radiation Laboratory and having served from 1963 to 1969 as the first associate director of the LLNL biology and medicine programme, he left the weapons laboratory in 1971 to resume his professorship at UC Berkeley. He became an emeritus professor in 1974 and continued to investigate the effects of low-level radiation on health well into his 80s.

3.3. Early Studies of Lipoproteins at the Donner Laboratory, 1949–1950

In 1949, Gofman and his colleagues in the Donner Laboratory of UC, Berkeley, described for the first time the quantification of low-density lipoproteins in the plasma of normal subjects by analytical ultracentrifugation, after first increasing the density of plasma by addition of sodium chloride, thereby enabling these lipoproteins to float rather than sediment during centrifugation.[4] The following year they identified by the same method an increase in lipoproteins with a Svedberg flotation rate (S_f) of 10–20 in the plasma of hypercholesterolaemic cholesterol-fed rabbits and human survivors of myocardial infarction, after first raising the density of plasma from 1.006 to 1.063. Gofman *et al.* observed that the concentration of lipoproteins in the S_f 10–20 class after ultracentrifugation correlated poorly with measurements of serum total cholesterol in myocardial infarct survivors and suggested that increases in low-density lipoproteins were a better marker of atherosclerosis than was a raised serum cholesterol.[5] The implications of this observation later generated a considerable amount of controversy in medical circles.

Fig. 3.1.   John Gofman (1918–2007) in 1979 (Photo: Egan O'Connor).

In 1955, after a further 5 years of research, Gofman and his colleagues published a lengthy account of their findings in the journal *Plasma.*[6] Unfortunately, the journal folded later that year, but happily, as described in what follows, Gofman's magnum opus was subsequently re-published in the *Journal of Clinical Lipidology* in 2007,[7] just a few months before his death at the age of 88 in San Francisco (Fig. 3.1).

3.4. Summary of Research on Lipoproteins at the Donner Laboratory, 1950–1955

3.4.1. *Separation and composition of lipoproteins in normal subjects*

In their detailed account of 6 years of work Gofman *et al.*[7] started by describing the methods they used to isolate plasma lipoproteins. The initial step involved using a preparative ultracentrifuge to separate the various classes of lipoproteins in serum according to their density. They adjusted the density of serum to 1.063, 1.125 and 1.20 g/ml in turn by the addition of sodium chloride, retrieved the lipoproteins from each supernatant after centrifugation and then quantified them by analytical ultracentrifugation. The latter procedure enabled Gofman to sub-divide lipoproteins of density <1.063, which they named "low-density lipoproteins," according to their flotation rate. The major constituent had an S_f of 0–12, with lesser amounts of lipoproteins of S_f 12–20, 20–100 and

100–400, the S_f value being roughly proportional to particle size. In modern terminology the S_f 0–12 class is synonymous with low density lipoprotein (LDL), the S_f 12–20 class with intermediate-density lipoprotein (IDL) and the S_f 20–400 class with very-low-density lipoproteins (VLDL). In a similar manner ultracentrifugation of serum at densities 1.125 and 1.21 yielded high-density lipoprotein (HDL)-2 and HDL-3, respectively. The lipids in the centrifugal sediment consisted mainly of non-esterified fatty acids bound to albumin.

Chemical analysis of extracted lipids showed that the chief lipid component of HDL was phospholipid, free and esterified cholesterol were the major lipid components of LDL and triglyceride was the predominant lipid in VLDL. The authors noted that the relative distribution of lipoproteins in individuals could differ greatly despite their having identical levels of serum cholesterol, reflecting differences in the lipid composition of the various lipoprotein classes. This observation inspired the distinction between the terms hyperlipidaemia and hyperlipoproteinaemia, introduced by Fredrickson *et al.*[8] in 1967 (see Chapter 7); the classification of lipoprotein phenotypes described in that review systematised the observations made a decade earlier by Gofman *et al.* in a variety of clinical situations.

3.4.2. *Physiological determinants of lipoprotein levels*

Detailed measurements of lipoproteins performed in hundreds of normal subjects showed that both age and gender had major influences on lipoprotein levels, with higher VLDL and LDL levels in males up to the age of 65 whereas HDL levels were higher in females at all ages up to 70. Marked increases in all the lipoprotein classes were observed in pregnancy, especially during the third trimester, but there was no concordance between lipoprotein levels in maternal and cord blood.

Detailed studies of the effect of intravenous heparin on lipoprotein levels in normal subjects suggested that VLDL is converted to LDL by a heparin-activated enzyme that hydrolyses triglyceride to free fatty acids. The enzyme responsible was identified in 1955 by Edward Korn at the US National Institutes of Health (NIH) and named lipoprotein lipase.[9] Prior to the advent of genetic analysis of lipoprotein lipase, absence of post-heparin lipolytic activity in plasma was used as an index of deficiency of the enzyme in patients with severe hypertriglyceridaemia.

3.4.3. *Analysis of lipoprotein changes in disease states*

Studies of 18 patients with xanthoma tendinosum (tumour-like accumulations of cholesterol in tendons), a feature of what nowadays is termed familial hypercholesterolaemia (FH), showed a marked increase in LDL, a small decrease in VLDL and marked decreases in both HDL-2 and HDL-3. An increase in LDL was also observed in patients with xanthelasma (xanthomas of the eyelid) but without xanthomas elsewhere. In contrast, patients with xanthoma tuberosum (cutaneous xanthomas), most of whom were probably suffering from what is now known as type III hyperlipoproteinaemia, showed marked increases in VLDL and IDL but marked decreases in LDL and HDL. Nine patients with essential hyperlipaemia, who exhibited lipaemic serum, had huge increases in lipoproteins in the S_f 20–400 range, which presumably included chylomicrons as well as VLDL, but marked decreases in LDL. Subsequent advances in research have revealed a primary, i.e. genetic, basis for most of these disorders.

Turning to secondary causes of lipoprotein abnormalities, Gofman *et al.* showed that patients with the nephrotic syndrome exhibited marked increases in both VLDL and LDL and marked decreases in HDL-2. Similar changes were observed in patients with chronic biliary obstruction. The lipoprotein changes found in patients with diabetic acidosis or coma were very similar to those observed in essential hyperlipaemia, with marked elevations of VLDL and reductions in LDL and HDL, and the extent of these changes depended upon how well the diabetic state was controlled. Marked elevations of LDL and IDL were observed in patients with hypothyroidism, which returned to normal after treatment with thyroid hormone. Subsequent research revealed that the latter regulates receptor-mediated catabolism of LDL.[10]

3.4.4. *Lipoprotein changes as predictors of coronary heart disease and atherosclerosis*

In previous studies, levels of both LDL and VLDL had been shown to be significantly higher in survivors of myocardial infarction than in control subjects. Subsequent studies suggested that VLDL was 1.6 times more potent than LDL in discriminating between these two groups. On the basis of this finding Gofman *et al.* developed a statistic termed α and calculated its value in the two groups. Mean values of α were higher in MI survivors, but differences lessened with increasing age. Hence, α alone did not predict risk of an MI, which is known to increase with age. To overcome this problem Gofman *et al.* developed an index that included both α and age, which they termed Accumulated Coronary Disease (ACD), and showed that ACD correlated with US coronary heart disease mortality rates at the time. The Donner group concluded their paper by stating that measurement of lipoprotein levels at any given age provides an estimate of the severity of coronary atherosclerosis, an assertion that generated considerable controversy at the time, as discussed in what follows.

In the report published by the Cooperative Study of Lipoproteins and Atherosclerosis in 1956, Gofman and his three Donner Laboratory colleagues described yet another Atherogenic Index that incorporated the combined values of S_f 12–20 and 20–400 lipoproteins and claimed that this discriminated better between patients with and without coronary heart disease than did measurements of serum cholesterol.[11] In contrast, the non-Donner members involved in the Cooperative Study, who comprised the majority, concluded that calculating the Atherogenic Index using complicated measurements of lipoproteins by analytical ultracentrifugation had no advantages over the much simpler measurement of serum cholesterol, an opinion shared by most clinicians.

Despite the criticisms, the data generated by the Donner Group using analytical ultracentrifugation were highly original and have greatly clarified our understanding of the pathophysiology of lipoprotein metabolism. Although Gofman's observations concerning the atherogenicity of S_f 12–20 and 20–400 particles were controversial at the time, they were also prophetic and antedated by 65 years current recommendations regarding the importance of triglyceride-rich lipoproteins (S_f 20–400) and their remnants (S_f 12–20) as risk factors.[12]

3.5. Richard Havel

There is a happy ending to this story. The practical problems posed by analytical ultracentrifugation, namely the expense, complexity and scarcity of analytical ultracentrifuges, were resolved by the publication in 1955 of a much simpler method of lipoprotein quantification by Havel *et al.* at the NIH.[13] This involved sequential preparative ultracentrifugation of serum in a Beckman Spinco 40.3 rotor, followed by extraction and chemical quantification of lipids in lipoproteins with densities of <1.019, 1.019–1.063 and >1.063, namely VLDL + IDL, LDL and HDL. The density of plasma was adjusted to d 1.019 and 1.063, respectively, with a mixture of sodium chloride and potassium bromide before sequential spins. Results in normal subjects showed that the concentration of lipoproteins of density <1.019 and 1.019–1.063 isolated by preparative ultracentrifugation correlated with those with flotation rates of S_f 20–400 and 0–12 quantified by Gofman *et al.* using analytical ultracentrifugation.

Havel's method greatly facilitated lipoprotein analysis and was the basis for the widely adopted "β quant" method of quantifying LDL cholesterol used in the Lipid Research Clinics Program. This involved preparative ultracentrifugation of plasma at its own density (d 1.006), removal of the supernatant by tube slicing followed by quantification of cholesterol in the d >1.006 infranatant. Subtraction from the latter of the value of HDL cholesterol, determined separately after heparin-manganese precipitation of plasma, provided an estimate of LDL cholesterol. In a tribute to Gofman published concomitantly with the re-publication in 2007 of the Donner Laboratory's findings, Havel acknowledged the seminal nature of Gofman's discoveries in the lipoprotein field.

3.6. Havel: Biography

Dick Havel's career started with a degree in Medicine plus a Master's degree in Chemistry at the University of Oregon Medical School in 1949. He did his residency in Internal Medicine at Cornell University Medical Center in New York and then began his research career at the National Heart Institute in Bethesda. He left there to join the faculty of Medicine at the University of California, San Francisco, in 1956 and was Director of the Cardiovascular Research Institute at UCSF from 1973 to 1992, where he trained many lipidologists. He was elected to the National Academy of Sciences of the USA in 1983 and edited the *Journal of Lipid Research* for several years.

Havel's paper on the isolation of lipoproteins from human serum by preparative ultracentrifugation is one the most cited publications in the field of lipid research[13] and enabled lipidologists to undertake a variety of research procedures, including studies of the turnover of lipoproteins in health and disease.[10] Together with his colleague Robert Gordon at the NIH, Havel was the first to show that an inherited deficiency of lipoprotein lipase was the cause of familial hyperchylomicronaemia.[14]

In conjunction with members of his UCSF Specialised Center of Research (SCOR) group, Havel undertook one of the first regression studies in patients with heterozygous FH.[15] This consisted of a randomised controlled trial in 72 men and women with FH, who underwent quantitative coronary angiography before and after 2 years of diet plus colestipol 15 g daily (controls) or of diet plus colestipol 20 g daily plus niacin up to 7.5 g daily, and in some instances plus lovastatin 40–60 mg daily (treatment group). Baseline LDL cholesterol was 7.3 mmol/l and averaged 6.3 and 4.5 mmol/l in the control and treatment groups, respectively, during the trial. Coronary lesions progressed in the former and regressed in the latter, demonstrating the benefit of lowering LDL cholesterol in FH.

This study was published in 1990, the year that Dick Havel and the author attended a lipid conference in South Africa, following which their host, David Marais, laid on a visit to the Kruger Park (Fig. 3.2).

Fig. 3.2.   Dick Havel (1925–2016) (left) with the author in the Kruger Park, South Africa in 1990.

Dick's slightly stiff manner and forced laugh on first acquaintance belied his warm nature and enthusiastic approach to life once one got to know him. He died in 2016 aged 91 after a long and extremely productive career in lipid research.[16]

Both Havel and Gofman exemplified the importance of a basic scientific background when undertaking high-quality clinical research. If Gofman was the Father of Clinical Lipidology, then Havel might well be regarded as a close relative.

References

1. Steinberg D. Anitschkow: Birth of the lipid hypothesis of atherosclerosis and coronary heart disease. In: *Pioneers of Medicine without a Nobel Prize*, Thompson G (ed). London: Imperial College Press; 2014.
2. Brown WV. Editorial. *J. Clin. Lipidol.* 2007; **1**: 97–99.
3. Havel RJ. Introduction: John Gofman and the early years at the Donner Laboratory. *J. Clin. Lipidol.* 2007; **1**: 100–103.
4. Gofman JW, Lindgren FT, Elliott H. Ultracentrifugal studies of lipoproteins of human serum. *J. Biol. Chem.* 1949; **179**: 973–979.
5. Gofman JW, Jones HB, Lindgren FT, Lyon TP, Elliott HA, Strisower B. Blood lipids and human atherosclerosis. *Circulation* 1950; **2**: 161–178.
6. Gofman JW, DeLalla O, Galzier F, *et al.* The serum lipoprotein transport system in health, metabolic disorders, atherosclerosis and coronary artery disease. *Plasma* 1955; **2**: 413–484.
7. Gofman JW, Delalla O, Glazier F, *et al.* The serum lipoprotein transport system in health, metabolic disorders, atherosclerosis and coronary heart disease. *J. Clin. Lipidol.* 2007; **1**: 104–141.
8. Fredrickson DS, Levy RI, Lees RS. Fat transport in lipoproteins — An integrated approach to mechanisms and disorders. *N. Engl. J. Med.* 1967; **276**: 34–42, 94–103, 148–156, 215–225, 273–281.
9. Korn ED. Clearing factor, a heparin-activated lipoprotein lipase I. Isolation and characterization of the enzyme from normal rat heart. *J. Biol. Chem.* 1955; **215**: 1–14.
10. Thompson GR, Soutar AK, Spengel FA, Jadhav A, Gavigan SJP, Myant NB. Defects of receptor-mediated low density lipoprotein catabolism in homozygous familial hypercholesterolemia and hypothyroidism *in vivo. Proc. Natl. Acad. Sci. USA* 1981; **78**: 2591–2595.
11. Cooperative Study of Lipoproteins and Atherosclerosis. Evaluation of serum lipoprotein and cholesterol measurements as predictors of clinical complications of atherosclerosis: Report of a cooperative study of lipoproteins and atherosclerosis. *Circulation* 1956; **14**: 691–742.

12. Ginsberg HN, Packard CJ, Chapman MJ, *et al*. Triglyceride-rich lipoproteins and their remnants: Metabolic insights, role in atherosclerotic cardiovascular disease, and emerging therapeutic strategies — A consensus statement from the European Atherosclerosis Society. *Eur. Heart J.* 2021; **42**: 4791–4806.
13. Havel RJ, Eder HA, Bragdon JH. The distribution and chemical composition of ultracentrifugally separated lipoproteins in human serum. *J. Clin. Invest.* 1955; **34**: 1345–1353.
14. Havel RJ, Gordon RS. Idiopathic hyperlipemia: Metabolic studies in an affected family. *J. Clin. Invest.* 1960; **39**: 1777–1790.
15. Kane JP, Malloy MJ, Ports TA, *et al*. Regression of coronary atherosclerosis during treatment of familial hypercholesterolemia with combined drug regimens. *JAMA* 1990; **264**: 3007–3012.
16. Kane JP, Malloy MJ. In memoriam: Richard J. Havel (1925–2016). *J. Lipid Res.* 2016; **57**: 1109–1110.

Chapter 4

Thomas Dawber and William Kannel: The Framingham Study and Risk Factors for Coronary Heart Disease

4.1. Introduction

The town of Framingham is situated in Massachusetts about 20 miles from Boston. Its main claim to fame is that it is the site of the most important and longest-running epidemiological study in the world to investigate the factors that predispose to the development of coronary heart disease and stroke. The stimulus to setting it up was the death of President Roosevelt in 1945 from hypertensive heart disease and stroke. This led to the allocation by the US Congress of $500,000 to fund a 20-year long observational study under the aegis of the US Public Health Service (PHS) aimed at (1) studying the expression of coronary artery disease in an unselected population and (2) determining factors responsible for its development by conducting clinical and laboratory examinations and long-term follow-up. The location of the study was influenced by the proximity of Harvard Medical School, a renowned centre of cardiological expertise.[1]

The Framingham Study was initially administered by Gilcin Meadows of the PHS, who started recruiting participants in 1948, but this function was soon taken over by the newly-formed National Heart Institute (NHI). The latter appointed Thomas Dawber as director of the study in 1950, a role he fulfilled for the subsequent 16 years. In 1966, sensing the threatened loss of NHI funding and closure of the study after its planned duration of 20 years had elapsed, Dawber moved to Boston University to raise $500,000 in private funds for its continuation and was succeeded as director by William Kannel. Dawber's financial ploy was so successful that in 1971 the NHI negotiated a federal contract with Boston University that safe-guarded future funding of the Framingham Study, which continues to this day more than 70 years after its inception. During that time three generations of Framingham residents have been recruited to the study: the Original cohort, starting in 1948; the Offspring cohort consisting of children of the Original cohort and their spouses, starting in 1971; and the Third Generation cohort, children of the Offspring cohort, starting in 2002. These, together with 1,000 or so additional recruits, totalled just over 15,000 persons who have participated in the study.[1]

4.2. Early Results of the Framingham Study

The first detailed results of the study were based on just under 4,500 Framingham residents, aged 30–59 in 1950, who had been followed up for 4 years by Dawber *et al.*[2] All had a detailed family history taken at entry and were examined by two physicians. Investigations included measurements of height and weight, chest X-Ray and ECG, and analyses of blood glucose and cholesterol. These procedures were repeated at biennial intervals for 4 years, during which 73 (1.6%) persons were known to have died and 28 (0.6%) lost to follow up. Arteriosclerotic heart disease (ASHD) was classified as myocardial infarction (MI), coronary occlusion (sudden death attributed to coronary heart disease (CHD)), evidence of MI solely on ECG ("silent"), angina pectoris and myocardial fibrosis.

Follow up data were available in 89% of participants. ASHD was found to be twice as prevalent in men as in women at entry to the study. In men it comprised equally angina and MI and it occurred mainly after the age of 40 in both sexes. All new ASHD events in women occurred after age 45, mainly angina, and in 80% of men. Almost half the heart attacks in men were fatal and one-third of these deaths were sudden. One-fifth of MIs were "silent." Hypertension (systolic blood pressure ≥160 mm Hg or diastolic ≥95 mm Hg) was strongly associated with new ASHD in men aged 45–62 as was obesity, the two being closely correlated. Hypercholesterolaemia (serum cholesterol ≥260 mg/dl or ≥6.7 mmol/l) was also strongly associated with new ASHD in men aged 45–62, but independently of hypertension and obesity.

These early results from Framingham during Dawber's directorship pointed to the importance of increasing age, male gender, hypertension and hypercholesterolaemia in predisposing to atherosclerotic cardiovascular disease. They set the stage for further exploration of remediable influences promoting this major cause of death and disability in the USA in the second half of the 20th century.

4.3. Dawber: Biography

Thomas "Roy" Dawber was born in 1913 in Canada where his father, a Methodist minister, and his mother had recently emigrated from England. The family moved to the USA when he was three and he later went to Haverford College in Pennsylvania and then on to Harvard Medical School, where he graduated in 1937. For the next 12 years he worked in the US Coast Guard service at Brighton Marine Hospital, ending up there as Chief of Medicine. He became the second director of the Framingham Study, publishing a detailed account of his experiences in his book

Fig. 4.1.   Thomas Dawber (1913–2005) (Photo: NHLBI Framingham Heart Study).

The Framingham Study.[3] In the preface he described the Study as an analysis of a chronic disease of unknown aetiology (Fig. 4.1).

He was appointed Chairman of Preventive Medicine and Epidemiology at Boston University in 1966, a position he held until he retired in 1980. He received numerous awards including the Gairdner Award in 1976 and was thrice nominated for the Nobel Prize. He was keen on carpentry, played several musical instruments and during his retirement spent much of his time sailing off the Florida coast. He died in Florida from Alzheimer's Disease in 2005, aged 92.[4]

4.4. Further Data from Framingham: Lipids as Risk Factors for Coronary Heart Disease

The word Framingham has become synonymous with the term "risk factors," inherited or acquired traits that increase the likelihood the bearer will develop coronary heart disease (CHD). The term was first coined in 1961 by Dawber's successor at Framingham, William Kannel.[5] In a subsequent report Kannel and his colleagues[6] focussed on the role of lipids as risk factors, specifically on the relationship between the concentrations of serum cholesterol and the major lipoprotein classes at entry to the study and development of CHD during the follow-up period.

The study population was relatively small, consisting of slightly more than 2,000 men and just under 3,000 women aged 30–62. All those with evidence of pre-existing CHD at the first examination were excluded and the remainder re-examined every 2 years. Serum cholesterol was measured on each occasion while the concentrations of S_f 0–20 and S_f 20–400 lipoproteins on analytical ultracentrifugation, corresponding to LDL and VLDL, respectively, were measured only at the first two examinations. Other measurements performed routinely were blood pressure, relative body weight, blood sugar, uric acid and smoking status. The occurrence of coronary events was based on ECG and enzyme changes and, if fatal, autopsy evidence. Of those enrolled, 80% were followed up for 14 years whereas an additional 8% died during that period.

The results showed that 323 men and 169 women developed CHD during the follow-up period, the incidence being threefold greater in those with serum total cholesterol and S_f 0–12 and S_f 20–400 lipoprotein concentrations in the top compared with the bottom quartile. However, except for S_f 20–400 lipoproteins in post-menopausal women, measurement of lipoprotein fractions was no more predictive of CHD than was serum cholesterol alone. The discovery that an increase in S_f 20–400 lipoproteins was a risk factor in older women, which manifests itself as a raised serum triglyceride, was a novel finding.

The risk associated with serum cholesterol was proportional to the latter's concentration, so that in men and women aged 35–44 the risk of CHD was fivefold greater if the serum cholesterol was >6.8 mmol/l compared with <5.7 mmol/l. The most striking example of this trend was the fate of six subjects found to have familial hypercholesterolaemia, all of whom had a serum cholesterol exceeding 10 mmol/l and died from CHD before the age of 50.

The Framingham Study not only established hypercholesterolaemia as a risk factor in its own right but also documented the multiplicative increase in risk that occurs when hypercholesterolaemia, hypertension and smoking all coexist.

In 1976, Kannel *et al.* expanded the list of risk factors to include glucose intolerance and left ventricular hypertrophy on ECG, which they incorporated with age, systolic blood pressure, serum cholesterol and smoking status into a logistic risk function.[7] The resulting data were used to estimate the probability by age and sex of developing cardiovascular disease (CVD) over the next 8 years of individuals free from it at the start. Men were at much greater risk than women up to the age of 70, after which the risk equalised. However, the following year the Framingham workers rendered these calculations obsolete by publishing data demonstrating the importance of the level of high density lipoprotein (HDL) cholesterol as an additional risk factor for CVD and acknowledging the need to include its value in calculations of risk.[8]

4.5. HDL Cholesterol and Risk of Cardiovascular Disease

It had long been known that HDL levels tended to be lower in situations associated with an increased risk of CHD, such as being male rather than female. In the 1950s, Gofman, using analytical ultracentrifugation (see Chapter 3), demonstrated an association between coronary artery disease and raised levels of the cholesterol carried in plasma by low density, but not by high density, lipoproteins. However, analytical ultracentrifugation was too expensive and cumbersome for routine use. Instead, Oliver and

Boyd, working in Edinburgh, utilised paper electrophoresis to separate α and β lipoproteins (corresponding to HDL and LDL, respectively) and showed that the α:β ratio was decreased in men with coronary disease, confirming Gofman's results.[9] However, it was not until the Miller brothers, based in Wales and Scotland, respectively, published a paper in *The Lancet* in 1975[10] that HDL really hit the scientific headlines.

Analysing published data from Finland, the Millers showed that HDL cholesterol levels were significantly lower in eight normocholesterolaemic patients with CHD than in 14 controls with similar levels of VLDL and LDL cholesterol. Again, using previously published data, they demonstrated an inverse correlation between HDL cholesterol concentration and the mass of cholesterol present in both the rapidly and slowly exchangeable pools of tissue cholesterol. Since the latter included arterial wall cholesterol, these authors proposed that having a low HDL cholesterol promotes the development of atherosclerosis by impairing the clearance of cholesterol from the arterial wall.

Miller and Miller noted that accumulation of tissue cholesterol is a feature of Tangier disease, a genetic disorder first described by Fredrickson *et al.* in which HDL is absent (see Chapter 7), and that HDL cholesterol is the preferred substrate of lecithin cholesterol acyltransferase (LCAT). The latter enzyme, together with cholesterol ester transfer protein (CETP), is involved in the movement of cholesterol from plasma to the liver, otherwise known as reverse cholesterol transport. They therefore concluded that strategies designed to raise HDL cholesterol would have a favourable effect on atherosclerosis by mobilising tissue cholesterol. However, the evidence for this has been decidedly limited so far, relying mainly on the equivocal results of trials using fibrates (see Chapter 6). Considering how little in the way of original data it contained, Miller and Miller's paper had an astonishing impact, being very highly cited in the literature. It undoubtedly reinforced acceptance of the inverse correlation between HDL cholesterol and risk of CHD, a relationship which was soon confirmed by data from the Framingham Study, as detailed in the following.[8]

Between 1969–1971, 2,815 men and women aged 49–62 in Framingham had their fasting levels of serum total, LDL and HDL cholesterol and triglycerides measured. Coronary heart disease developed in 79 men and 63 women over the course of 4 years. The statistically most significant lipid risk factor was HDL cholesterol, which was inversely correlated with CHD. Persons with a low HDL cholesterol (<35 mg/dl or <0.9 mmol/l) had an incidence of CHD that was eight times greater than those with a high HDL cholesterol (>65 mg/dl or >1.7 mmol/l). There was a weaker positive correlation between LDL cholesterol and CHD.

On univariate analysis HDL cholesterol was negatively correlated with all manifestations of CHD except CHD death in men. It was also negatively correlated with triglycerides and relative body-weight, but the Framingham workers rejected the proposition that the risk conferred by having a low HDL cholesterol reflected its correlation with a raised serum triglyceride.[11] However, so far there has been no hard evidence that a low HDL cholesterol is causally related to CHD. Proof of causality would require demonstrating that raising the level of HDL by therapeutic means reduced the risk, unequivocal evidence of which is currently lacking. Nevertheless, HDL cholesterol is now accepted as an important index of risk of CHD and is integral to all risk factor calculations, including the algorithm predicting the 30-year risk of CHD events and stroke based on data from the Framingham Offspring Study.[12]

4.6. Kannel: Biography

William "Bill" Kannel spent much of his career in the Framingham Study, of which he was a major driving force and which he directed from 1966–1979. He was born in Brooklyn, New York, in 1923 and obtained his medical degree at the Medical College of Georgia in Augusta. He did his training in internal medicine with the US Public Health Service on Staten Island, nowadays the site of the start of the New York Marathon.

Kannel joined the National Heart Institute in 1951 to work in the Framingham Study and continued to do so after becoming Professor of Medicine at Boston University in 1979. He received numerous awards during his career including the Gairdner International Award in 1976 concomitantly with Dawber, "for their careful epidemiologic studies, revealing risk factors with important implications for the prevention of cardiovascular disease." The Gairdner Award is regarded as the Canadian equivalent of the Nobel Prize.

Peter Wilson, one of Kannel's protégés, invited the author to Framingham to meet Kannel (Fig. 4.2), who kindly agreed to write the Foreword to a book co-authored by the author and Wilson, *Coronary Risk Factors and their Assessment*.[13] Peter Wilson subsequently left Framingham and was appointed Professor of Medicine in the Division of Cardiology at Emory University School of Medicine, Atlanta.

Fig. 4.2.   William Kannel (1923–2011).

Kannel died in 2011 in Massachusetts, aged 87. In his obituary he was described as an inspirational driving force and an enormous contributor to the prevention of cardiovascular disease as well as being the founding father of several generations of researchers who have made and continue to make important contributions in preventive cardiology.[14]

4.7. Past Achievements of Framingham and Future Research

Triggered by the death of President Roosevelt from hypertension after the Second World War and spurred on by the extremely high mortality from cardiovascular disease in the USA in the middle of the 20th century, the newly formed National Heart Institute set up the Framingham Heart Study to investigate factors that predisposed to heart disease. While the number of Framingham residents recruited was relatively modest, the Study has generated an enormous quantity of scientific data over many years. Although epidemiological data from observational studies cannot prove causality, nevertheless they provide strong clues as to which directions further research should take. Hence, the well-documented association between LDL and cardiovascular disease was eventually shown to be causal by the decrease in coronary morbidity and mortality that resulted from lowering LDL cholesterol by statin therapy (see Chapter 17).

Despite the lack of evidence that therapeutic elevation of HDL is beneficial, the plasma level of HDL cholesterol continues to be regarded as an important index of risk. One confounding influence is the strong inverse relationship between HDL and triglyceride-rich lipoproteins such as VLDL and IDL and it has been suggested that these are just as atherogenic as LDL.[15] Thus, it could be that HDL cholesterol levels are not causally associated with cardiovascular disease but simply reflect inverse changes in triglycerides. Alternatively, it may be that HDL has cholesterol-mobilising effects in its own right, as suggested by Miller and Miller,[10] or possesses an anti-atherosclerotic action related to its anti-inflammatory properties.[16]

In addition to confirming Anitschkow's and Gofman's research on the importance of lipid abnormalities as a risk factor, the Framingham Study established the role of raised blood pressure, especially systolic, as a major risk factor for coronary heart disease and stroke. Hypertension, hypercholesterolaemia and cigarette smoking remain the main global determinants of cardiovascular disease, although the prevalence of all three has decreased considerably in the USA during the Study's long tenure. To what extent this is cause and effect is impossible to say, but Framingham's findings have undoubtedly influenced both public health policy and clinicians' attitudes.

An informative and up-to-date summary of the structure and achievements of the Framingham Study and the directions which its future research is taking was published recently by Framingham scientists.[17] This includes studying temporal trends in risk factors for CVD, a purpose for which the Study is ideally suited because of its standardised measurements of risk factors and stringent criteria for assessing outcomes. Secondly, it has expanded its range of investigations to include cardiopulmonary exercise testing, with the aim of improving understanding of impaired exercise tolerance in, for example, hypertensive subjects with heart failure. And thirdly, its transgenerational design makes it ideal for genetic and genomic studies, as does its storage of more than 1.5 million samples that are available for future use in collaborative research projects. The Framingham Study may be over 70 years old, but there's plenty of life left in it.

References

1. Mahmood SS, Levy D, Vasan RS, Wang TJ. The Framingham Heart Study and the epidemiology of cardiovascular diseases: A historical perspective. *Lancet* 2014; **383**: 999–1008.
2. Dawber TR, Moore FE, Mann II. GV. Coronary Heart Disease in the Framingham Study. *Am. J. Public Health Nations Health.* 1957; **47**: 4–24.
3. Dawber TR. *The Framingham Study.* Cambridge, MA: Harvard University Press; 1980.
4. Obituaries: Thomas Royle Dawber. *BMJ* 2006; **332**: 122.
5. Kannel WB, Dawber TR, Kagan A, Revotskie N, Stokes J 3rd. Factors of risk in the development of coronary heart disease — Six year follow-up experience. The Framingham Study. *Ann. Intern. Med.* 1961; **55**: 33–50.
6. Kannel WB, Castelli WP, Gordon T, McNamara PM. Serum cholesterol, lipoproteins, and the risk of coronary heart disease. The Framingham study. *Ann. Intern. Med.* 1971; **74**: 1–12.
7. Kannel WB, McGee D, Gordon T. A general cardiovascular risk profile: The Framingham Study. *Am. J. Cardiol.* 1976; **38**: 46–51.
8. Gordon T, Castelli WP, Hjortland MC, Kannel WB, Dawber TR. High density lipoprotein as a protective factor against coronary heart disease. *Am. J. Med.* 1977; **62**: 707–714.
9. Oliver MF, Boyd GS. Serum lipoprotein patterns on coronary sclerosis and associated conditions. *Br. Heart J.* 1955; **17**: 299–302.
10. Miller GJ, Miller NE. Plasma-high-density-lipoprotein concentration and development of ischaemic heart disease. *Lancet* 1975; **1**: 16–19.
11. Carlson LA, Ericsson M. Quantitative and qualitative serum lipoprotein analysis. Part 2. Studies in male survivors of myocardial infarction. *Atherosclerosis* 1975; **21**: 435–450.
12. Pencina MJ, D'Agostino RB Snr, Larson MG, Massaro JM, Vasan RS. Predicting the 30-year risk of cardiovascular disease: The Framingham Heart Study. *Circulation* 2009; **119**: 3078–3084.
13. Thompson GR, Wilson PW. *Coronary Risk Factors and their Assessment.* London, UK: Science Press; 1992.
14. In memoriam William B. Kannel 1923–2011. *Texas Heart Inst. J.* 2011; **38**: 615–616.
15. Nordestgaard BG, Varbo A. Triglycerides and cardiovascular disease. *Lancet* 2014; **384**: 626–635.
16. Jia C, Anderson JL, Gruppen EG, *et al.* High-density lipoprotein anti-inflammatory capacity and incident cardiovascular events. *Circulation* 2021; **143**: 1935–1945.
17. Andersson C, Nayor M, Tsao CW, Levy D, Vasan RS. Framingham Heart Study: *JACC* Focus Seminar, 1/8. *J. Am. Coll. Cardiol.* 2021; **77**: 2680–2692.

Chapter 5

Ancel Keys: The Seven Countries Study and the Role of Dietary Fat in Determining Cholesterol Levels in Populations

5.1. Introduction

The studies described in the preceding chapters established the role of lipids as risk factors but did little to elicit the underlying causes of the lipid abnormalities. A possible exception was the identification in the Framingham Study of a small number of subjects with familial hypercholesterolaemia (FH), a relatively rare condition known to be genetically determined and associated with a very high risk of CHD. However, most subjects at increased risk of CHD in Framingham had much lower levels of cholesterol than those seen in FH, albeit higher than those of subjects who were at low risk of CHD. An alternative explanation for the modest elevations of serum cholesterol seen in the majority of persons at high risk was that it had an environmental cause, such as diet.

This topic became the object of intense research by the American physiologist Ancel Keys, whose findings in his Seven Countries Study generated a considerable amount of controversy when they were published in 1970 and long afterwards. For example, in 1977 George Mann, a career investigator of the NIH, accused Keys of deliberately selecting data from the literature which showed that coronary heart disease was correlated with dietary fat intake. He claimed instead that the Seven Countries Study demonstrated that differences in the frequency of coronary disease between countries, based on ECG evidence, were correlated with differing amounts of occupational activity, not with a high fat diet, and that a raised serum cholesterol simply reflected a sedentary life-style.[1] These criticisms were later vigorously rebutted by Keys's colleague Henry Blackburn, who accused Mann of failing to distinguish between personal attacks on colleagues and scientific criticism and claimed that very few scientists challenged the relationship between diet, blood lipoproteins and risk of coronary disease.[2]

5.2. Background

Ancel Keys was a physiologist at the University of Minnesota with a strong interest in the epidemiology of coronary heart disease (CHD). In 1947, he initiated a small prospective study of this disorder in a group of 281 local businessmen and professionals, who he followed up for 15 years.[3] During that time definite CHD developed in 32 (11.4%) of the men and possible CHD in 16 (5.7%). The incidence of CHD was strongly correlated with serum cholesterol and to a lesser extent with systolic blood pressure, and those who developed CHD tended to have low levels of α lipoprotein (HDL). The authors, including Blackburn, commented that their data showed a high degree of concordance with the data emerging from Framingham.

In 1957, Keys *et al.* published data from studies in men on diets that varied in fat content and composition and showed that saturated fats raised serum cholesterol and that polyunsaturated fats lowered it. Furthermore, they found that the cholesterol-raising effect of saturated fat was twice as great as the cholesterol-lowering effect of an equivalent amount of polyunsaturated fat, e.g. +10% vs −5%.[4] Exploratory surveys he conducted suggested to Keys that differences in dietary fat might be responsible for geographical differences in serum cholesterol and in the prevalence of CHD between countries such as South Africa, where it was rare in indigenous black inhabitants, and the USA where 40% of deaths in white middle-aged men were caused by CHD. To pursue this question further, in 1956 Keys undertook feasibility studies in countries with the lowest and highest recorded rates of CHD in the world, namely Japan and Finland. His findings confirmed the feasibility of conducting the Seven Countries Study, which started the following year supported by grants from the US Public Health Service, American Heart Association and several other organisations.

5.3. The Seven Countries Study

The Seven Countries Study formally started in 1958 and between then and 1964 a total of 16 cohorts amounting to 12,270 men aged 40–59 were enrolled in seven countries across four regions of the world (United States, Northern Europe, Southern Europe, Japan). The seven countries involved in the Study were Finland, Greece, Italy, Japan, the Netherlands, the USA and Yugoslavia. The objective was to examine all men in defined areas of each country at five-yearly intervals over a period lasting at least 10 years. All the participants were examined using standardised methods, each examination including a questionnaire, blood and urine analysis, and electrocardiograms at rest and after exercise. Dietary intake was assessed by random surveys, which involved weighing all food consumed over 1 week and chemical analysis of replicate meals. Dietary surveys were repeated at different times of the year. All deaths and major illnesses were recorded and complete five-year examinations were performed in 94% of survivors.

In 1970, an entire issue of *Circulation Supplement* was devoted to 20 articles that described the design and five-year results of the Seven Countries Study.[5] The results confirmed that the age-standardised prevalence of CHD at entry was much higher in Finland and the USA than in most other countries, including evidence of past myocardial infarction on ECG. Subjects with evidence of a previous myocardial infarction at entry to the study had a death rate almost 14 times greater than the remainder. CHD prevalence rates were related to blood pressure and serum cholesterol but surprisingly not to relative body weight or smoking habits.

During the next 5 years the incidence of CHD in men initially free of CHD followed a similar geographical pattern, being highest in Finland, especially East Finland where it occurred at twice the rate of West Finland, next highest in the USA, and lowest in Japan and Greece. The incidence of CHD was twice as high in men with an abnormal exercise ECG at the start of the study as in those who were initially free from CHD. There were 588 deaths over the course of the 5 years, 158 of them due to CHD. Overall, the proportion of total deaths due to CHD was only 12.5% but it was much higher in the Finnish and US cohorts, reaching 34% and almost 50%, respectively.

Differences in the incidence of CHD between countries could not be explained by differences in obesity or smoking but were closely correlated with differences in the frequency of hypercholesterolaemia and with the percentage of calories provided by saturated fat in each country. For example, 56% of Finnish men had a serum cholesterol of >250 mg/dl (6.5 mmol/l) and 20% of their calories were provided by dietary saturated fat compared with 7% and 3%, respectively, in Japan, with comparable differences in the incidence of CHD between the two countries. In the light of these findings, Keys lobbied health organisations and government bodies and subsequently achieved major dietary changes in the USA, Finland and most other Western nations, resulting in a reduction in saturated fat consumption and its partial replacement with polyunsaturated fat.

5.4. Biography

Ancel Keys was born in Colorado in 1904 and grew up in California, where his uncle Lon Chaney was a star of silent horror films. Keys must have been a bit of a tearaway in his youth because the jobs he did included shovelling bat guano in caves in Arizona, working in a lumber camp and also as a powder monkey in a goldmine in Colorado. Subsequently, he started but soon abandoned a degree course in Chemistry at the University of California at Berkeley in order to work as a greaser in the engine room of a ship to China. On his return, he resumed his studies at Berkeley to achieve a BA in Economic and Political Science and an MSc in Zoology. In 1930, he got a PhD in Oceanography and Biology at the Scripps Institute of Oceanography in San Diego, followed by a PhD in Physiology at King's College, Cambridge. He returned to the States to work in the Fatigue Laboratory at Harvard University on the physiological effects of high altitude, which included his spending 10 days at 20,000 feet in the Andes. In 1937, he was invited to set up the Laboratory of Physiological Hygiene at the University of Minnesota, which he directed and where he conducted research into the health benefits of physical activity and proper nutrition until his retirement in 1972 (Fig. 5.1).

During the Second World War Keys was asked by the US Government to devise a compact food pack for paratroopers, which soon became famous as the K ration. After the war he was appointed a Senior Fulbright Fellow at Oxford and in 1950 chaired a WHO Commission on Food and Agriculture in Rome. Keys recognised that there had been a dramatic fall

Fig. 5.1. Ancel Keys (1904–2004) (Reprinted from Oransky I. Ancel Keys. *The Lancet* 2004; 364: 2174, with permission from Elsevier).

in heart attacks in parts of post-war Europe where starvation was rife and this stimulated him to embark on the Seven Countries Study. Known as "Mr Cholesterol," his work made him famous and he appeared on the cover of *Time* magazine in January 1961 under the banner heading "Diet and Health."[6]

After he retired he went to live in Southern Italy and helped to popularise the benefits of the olive oil-based Mediterranean diet, now a corner stone of healthy-lifestyle strategies to prevent CHD. To mark his 100th birthday two of the major figures in preventive cardiology, Mario Mancini and Jeremiah Stamler, co-authored a warm tribute to Keys's pioneering research into the role of diet in cardiovascular disease.[7] This was published shortly before he died at his home in Minneapolis in 2004, aged 100.

Stamler himself was the organiser of the Multiple Risk Factor Intervention Trial (MRFIT), which established a graded and continuous relationship between serum cholesterol and CHD. He died in 2022 aged 102 and was described in his obituary as the "Father of Preventive Cardiology."[8]

5.5. National Diet–Heart Feasibility Study

The results of the Seven Countries Study reinforced data from other sources that suggested that increased levels of serum cholesterol, especially LDL cholesterol, were associated with the prevalence of CHD but evidence that this relationship was causal was lacking. In 1960, an Executive Committee, of which Ancel Keys was a member, was set up by the National Heart Institute to examine the feasibility of a National Diet–Heart Study aimed at investigating the effect of dietary change on the occurrence of clinical CHD in the United States.[9] The Committee recognised that a mass field trial would necessitate following up 100,000 free-living men for 4–5 years and would be a formidable undertaking. Its final report concluded that such a trial could probably only be undertaken on a non-double-blind basis.[10] Another consideration must have been the enormous cost of such a trial.

An alternative approach was to undertake a double-blind study in institutionalised populations, where better-supervised dietary control and therefore greater reductions in serum cholesterol, together with better documentation of events, would reduce the number of individuals needed to achieve a statistically significant outcome. A consequence of this conclusion was the NIH funded Minnesota Coronary Survey,[11] with Ivan Frantz, Professor of Medicine and Biochemistry at the University of Minnesota, as principal investigator and Ancel Keys as co-principal investigator.

5.6. The Minnesota Coronary Survey

The Minnesota Coronary Survey was a double-blind, randomised clinical trial conducted in six Minnesota state mental hospitals and a nursing home between 1968 and 1973. It involved 4,393 institutionalised men and 4,664 institutionalised women. Nowadays such a trial would be prohibited for ethical reasons, but 50 years ago it was considered acceptable.

The effects of a treatment diet containing 38% total fat (9% saturated fat, 15% polyunsaturated fat, 14% monounsaturated fat) and 166 mg dietary cholesterol per day were compared with those of a control diet containing 39% total fat (18% saturated fat, 5% polyunsaturated fat, 16% monounsaturated fat) and 446 mg dietary cholesterol per day on serum cholesterol levels and on the incidence of myocardial infarctions, sudden deaths and all-cause mortality. The mean duration of time on the diets was 384 days, but only 17% of subjects consumed the diet for over 2 years.

The mean serum cholesterol level at entry to the trial was 207 mg/dl (5.3 mmol/l), decreasing to 175 mg/dl (4.5 mmol/l) in the treatment group (−15%) and to 203 mg/dl (5.2 mmol/l) in the control group (−2%), but no significant differences were observed between the treatment and control groups for cardiovascular events, cardiovascular deaths or total mortality. There was a trend towards fewer cardiovascular events in men and women aged 35–55 who had been on the treatment diet for >2 years, but the number of individuals was very small. In contrast, all-cause mortality over the course of four and a half years tended to be higher among those of all ages who were in the treatment group.

The distinctly negative results of this trial must have been very disappointing for Ancel Keys and his name was not on the publication that belatedly followed in 1989. This helped shield him from the adverse publicity concerning dietary fat and CHD that resulted at the time. However, he had already died before the storm really broke, following the publication of previously unpublished data that Ivan Frantz's son Robert unearthed from his father's basement after the latter's death in 2009, as described in what follows.

5.7. Analysis of Recovered Data from the Minnesota Coronary Survey

In 2016, Christopher Ramsden, working in the Section on Nutritional Neurosciences, National Institute on Alcohol Abuse and Alcoholism at the NIH, and his colleagues published an analysis of previously unpublished data from the Minnesota Coronary Survey (or Experiment as they renamed it) discovered by Robert Frantz. They also included a meta-analysis of published cholesterol-lowering trials that involved substituting linoleic acid-rich polyunsaturated fat for saturated fat.[12]

The data Frantz's son unearthed enabled Ramsden *et al.* to undertake a complete analysis of changes in serum cholesterol in the 2,355 participants exposed to the study diets for a year or more and also to analyse 149 autopsy records. They found that the intervention group had a significant reduction in serum cholesterol compared with controls (mean change from baseline −13.8% vs. −1.0%; $P < 0.001$) but there was no evidence of a reduction in coronary atherosclerosis or myocardial infarcts in the intervention group nor in mortality. Instead, there was a 22% higher risk of death for each 30 mg/dL (0.78 mmol/L) reduction in serum cholesterol

($P < 0.001$). In addition, their meta-analysis of five randomized controlled trials of similar cholesterol-lowering diets ($n = 10,808$) confirmed that the latter did not reduce mortality from coronary heart disease.

Ramsden *et al.* concluded that replacement of saturated fat in the diet with linoleic acid effectively lowers serum cholesterol but that the data did not support the hypothesis that this reduces risk of coronary or total mortality. They further suggested that the findings of the Minnesota Coronary Experiment added to growing evidence that incomplete publication of trial data contributed to overestimation of the benefits of replacing saturated fat in the diet with polyunsaturated fats rich in linoleic acid. The fact that they published their paper in the *British Medical Journal* could reflect difficulty in publishing it in an American journal. Whether or not this was so, the story captured the US public's imagination and led to headlines such as "Records found in dusty basement undermine decades of dietary advice" in *Scientific American* and "A decade-old study challenges advice on saturated fat" in the *New York Times*.

5.8. Official Statements on Dietary Fat Intake

As long ago as 1957, the American Heart Association (AHA) had published advice to the American public aimed at modifying the national diet with the objective of decreasing the prevalence of CHD.[13] Further policy statements were issued every 3–5 years from then until 1988. In 1989, representatives of nine health organisations and government bodies met under the aegis of the AHA and after reviewing the scientific evidence concluded that most Americans could improve their overall health by reducing consumption of fat, especially saturated fat, and cholesterol, increasing consumption of complex carbohydrates and dietary fibre and achieving an appropriate body weight.[14]

In 1996, the AHA specified that the nation should consume 30% or less of total calories/day from fat, of which 8–10% is saturated fat, up to 10% polyunsaturated fat, up to 15% monounsaturated fat and less than 300 mg/day of cholesterol, with complex carbohydrates providing 55%–60% of the total calories consumed.[11] Similar advice had been issued in the UK by the Committee on Medical Aspects of Food Policy (COMA), which also promoted increasing the consumption of foods rich in ω-3 fatty acids.[15]

A year after the publication of Ramsden *et al.*'s re-evaluation of the traditional diet–heart hypothesis, the AHA published its own very detailed analysis of the relationship between diet and CHD.[16] This included a meta-analysis of 4 so-called core trials of lowering cholesterol by substituting polyunsaturated (P) for saturated fat (S), which showed a 29% reduction in CHD events in those on a high P/S ratio diet. Ramsden *et al.*'s re-analysis of the participants who remained in the Minnesota Coronary Survey for at least 1 year was excluded from the AHA's meta-analysis because of its short duration, large percentage of withdrawals, intermittent exposure to treatment, and the inclusion of lightly hydrogenated corn oil margarine in the polyunsaturated fat diet. This type of margarine contains *trans* linoleic acid, a type of *trans* fatty acid strongly associated with increased risk of CHD. Hence, the AHA was unconvinced by the criticisms voiced by Ramsden *et al.* and maintained its long-standing recommendations to replace saturated fat with polyunsaturated and monounsaturated fat to lower the incidence of CVD in the US population.

5.9. Temporal Changes in Diet, Risk Factors and CHD in the USA

Age-adjusted death rates from CHD in US adults aged 25–84 fell from 543 to 267/100,000 in men (–51%) and from 263 to 134/100,000 in women (–49%) between 1980 and 2000 and similar downward trends occurred in the UK (Fig. 5.2). Roughly half of the reduction in US deaths was attributed to changes in risk factors, including decreases in serum cholesterol (–24%), systolic blood pressure (–20%) and smoking (–12%).[17]

National Health and Nutrition Examination Surveys (NHANES) conducted between 1971–1975 (NHANES 1) and 2005–2006 (NHANES 2005–2006) indicate that significant changes in the US diet took place during the interim: total fat intake expressed as percent of energy fell from 36.5% to 33.2% in women and from 37% to 33.3% in men; the consumption of saturated fat in both sexes fell from 13.4% to 11.6% of total energy and the dietary intake of cholesterol fell from 394 to 286 mg/day; data on mono- and polyunsaturated fat consumption were not available during NHANES 1.[18]

Data from the Minnesota Heart Survey showed that between 1980–1982 and 2000–2002 serum cholesterol decreased in US men and women by 6% and 5.4%, respectively.[19] Although there was a marked

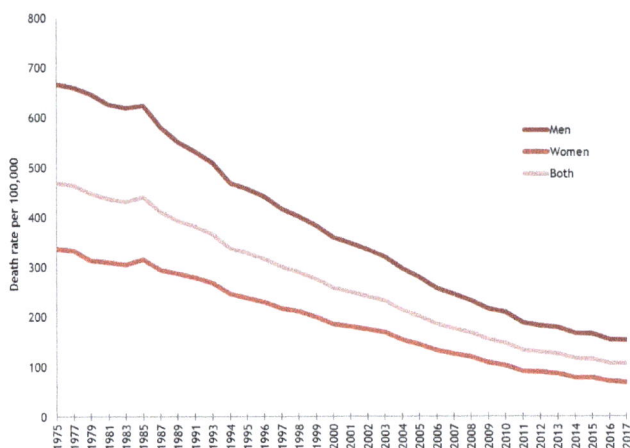

Fig. 5.2. Age-standardised death rate per 100,000 from coronary heart disease in the UK, 1975–2017 (Reproduced from British Heart Foundation (BHF) coronary heart disease statistics at bhf.org.uk/statistics). The decrease in CHD deaths was established well before statin use became common (see Fig. 5.3).

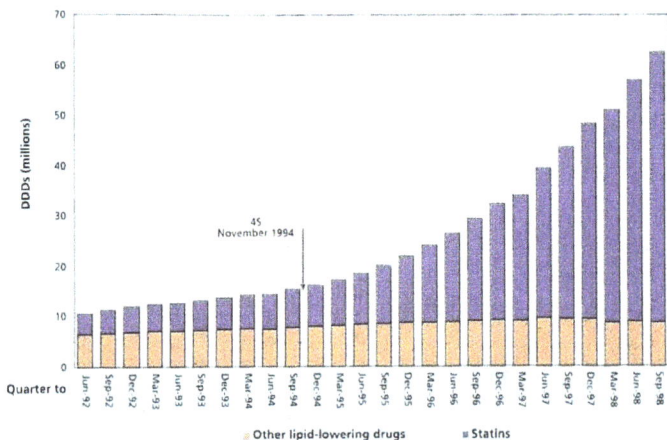

Fig. 5.3. Exponential increase in prescribing of statins in England before and following publication of the results of 4S in 1994 (DDD, defined daily dose).

increase in statin therapy in the USA and UK following the publication of 4S in 1994 (Fig. 5.3), the decreases in saturated fat and cholesterol consumption documented by NHANES must have contributed to the decrease in serum cholesterol and therefore to the 50% reduction in CHD observed between 1980 and 2000 in the USA. These data helped to vindicate the work of Keys and his colleagues and refute critics of the diet–heart hypothesis he espoused half a century ago, although there are still some who question whether saturated fat is harmful.[20]

References

1. Mann GV. Diet-Heart: End of an era. *N. Eng. J. Med.* 1977; **297**: 644–650.
2. Blackburn H. George Mann's editorial on "Diet-Heart: End of an era?" A reply. http://www.epi.umn.edu/cvdepi/essay/george-manns-editorial-on-diet-heart-end-of-an-era-a-reply/.
3. Keys A, Taylor HL, Blackburn H, *et al.* Coronary heart disease among Minnesota business and professional men followed 15 years. *Circulation* 1963; **28**: 381–395.
4. Keys A, Anderson JT, Grande F. Prediction of serum-cholesterol responses of man to changes in fats in the diet. *Lancet* 1957; **273**: 959–966.
5. Coronary Heart Disease in seven countries. *Circulation* 1970; **41**(Supplement 1): I-1-198.
6. Obituary: Ancel keys. November 26th 2004, *The Times.*
7. Mancini M, Stamler J. Diet for preventing cardiovascular diseases: Light from Ancel Keys, distinguished centenarian scientist. *Nutr. Metab. Cardiovasc. Dis.* 2004; **14**: 52–57.
8. Gotto AM, Olsson AG. Professor Jeremiah Stamler (1919–2022), "Father of Preventive Cardiology." *Atherosclerosis* 2022 March 12; S0021-9150(22) 00110-1. doi: 10.1016/j.atherosclerosis.2022.03.001.
9. Baker BM, Frantz ID Jr, Keys A. The national diet-heart study. *JAMA* 1963; **185**: 105–106.
10. The national diet-heart study final report. *Circulation* 1968; **37**(3 Suppl): I1-428.
11. Frantz ID Jr, Dawson EA, Ashman PL, *et al.* Test of effect of lipid lowering by diet on cardiovascular risk. The Minnesota Coronary Survey. *Arteriosclerosis* 1989; **9**: 129–135.
12. Ramsden CE, Zamora D, Majchrzak-Hong S, *et al.* Re-evaluation of the traditional diet-heart hypothesis: Analysis of recovered data from Minnesota Coronary Experiment (1968–1973). *BMJ* 2016; **353**: i1246. doi: 10.1136/bmj.i1246.
13. Krauss RM, Deckelbaum RJ, Ernst N, *et al.* Dietary guidelines for healthy American adults. A statement for health professionals from the nutrition committee, American Heart Association. *Circulation* 1996; **94**: 1795–1800.

14. American Heart Association. The healthy American diet. *Circulation* 1991; **82**: 1079.
15. Committee on Medical Aspects of Food Policy. Nutritional aspects of cardiovascular disease. London: HMSO; 1994.
16. Sacks FM, Lichtenstein AH, Wu JHY, *et al.* Dietary fats and cardiovascular disease: A presidential advisory from the American Heart Association. *Circulation* 2017; **136**: e1–e23. doi: 10.1161/CIR.0000000000000510.
17. Ford ES, Ajani UA, Croft JB, *et al.* Explaining the decrease in US deaths from coronary disease, 1980–2000. *N. Engl. J. Med.* 2007; **356**: 2388–2398.
18. Austin GL, Ogden LG, Hill JO. Trends in carbohydrate, fat, and protein intakes and association with energy intake in normal-weight, overweight, and obese individuals: 1971–2006. *Am. J. Clin. Nutr.* 2011; **93**: 836–843.
19. Arnett DK, Jacobs DR, Luepker RV, *et al.* Twenty-year trends in serum cholesterol, hypercholesterolemia, and cholesterol medication use: The Minnesota heart survey, 1980–1982 to 2000–2002. *Circulation* 2005; **112**: 3884–3891.
20. Heileson JL. Dietary saturated fat and heart disease: A narrative review. *Nutr. Rev.* 2020; **78**: 474–485.

Chapter 6

Michael Oliver: Pros and Cons of Lowering Cholesterol

6.1. Introduction

Apart from Anitschkow, the previous chapters have focussed exclusively on research conducted by American scientists. Hence, it is appropriate now to consider the research being undertaken in the 1950s on the other side of the Atlantic. The chief personality involved in research on lipids in Britain in those days was Michael Oliver, a charismatic cardiologist with the rare attribute of being interested in the role of cholesterol in atherosclerosis. He had an extremely successful career as a cardiologist whereas his research into lipids in relation to coronary heart disease had an ambivalent quality, seesawing between supporting and then questioning the treatment of hyperlipidaemia, opinions which he freely expressed both in speech and in print.

6.2. Biography

Michael Oliver was born in 1925 in Wales and was educated at Marlborough College. After the death of his father, a retired soldier, his mother married a doctor who influenced him to study Medicine. He qualified at Edinburgh University in 1947 and subsequently trained as a cardiologist at the Edinburgh Royal Infirmary. Together with Desmond Julian, he set up a coronary care unit there, the first in Europe. He was appointed to the Duke of Edinburgh Chair of Cardiology at the Royal Infirmary in

Edinburgh in 1976, elected President of the British Cardiac Society in 1980 and was the recipient of several honorary degrees and fellowships.

In 1957, while working in Edinburgh, he had written to the pathologist John French at the Dunn School of Pathology in Oxford, expressing concern that there was no organisation in Britain interested in the relationship of lipid metabolism, blood flow and platelets to experimental atherosclerosis. French invited Oliver to come to Oxford to meet his boss, Sir Howard Florey, who, having won the Nobel Prize for his work on penicillin, had now become interested in vascular pathology. At their meeting Florey asked French to make a list of those working in the field of atherosclerosis and asked Oliver to raise enough money to hold an inaugural meeting. With the help of a friendly contact at Glaxo, "Cuth" Cuthbertson, Oliver raised the required £250, and French produced a list of people to invite.

The inaugural meeting of the new group, comprising 27 members, was held in the autumn of 1958 at the old MRC headquarters in London, the topic being "Interrelationships of dietary fat, blood lipids and atherosclerosis." Lord Florey, as he became, was the first chairman and Michael Oliver the first secretary of the Atherosclerosis Discussion Group, later renamed the British Atherosclerosis Society.

Despite his Scottish affiliations, Michael Oliver remained an archetypical Englishman throughout his life. He retired in 1989 to live in London and died in 2015 in Umbria, Italy, where he had a second home, aged 89.

6.3. Wellcome Trust Witness Seminar

In 2005, Michael Oliver helped to organise and chair a Wellcome Trust Witness Seminar on "Cholesterol, Atherosclerosis and Coronary Disease in the UK, 1950–2000," which took place in London and in which the author participated[1] (Fig. 6.1). In his introductory address, Oliver stated that the purpose of the meeting was to record what scientists in the UK were thinking about cholesterol and associated lipids, and their relation to atherosclerosis, particularly coronary heart disease, in the last half of the 20th Century. His own thoughts on this topic are paraphrased in what follows.

In the beginning of the 1940s, coronary heart disease was regarded as an inevitable result of ageing. For example, in a monograph on

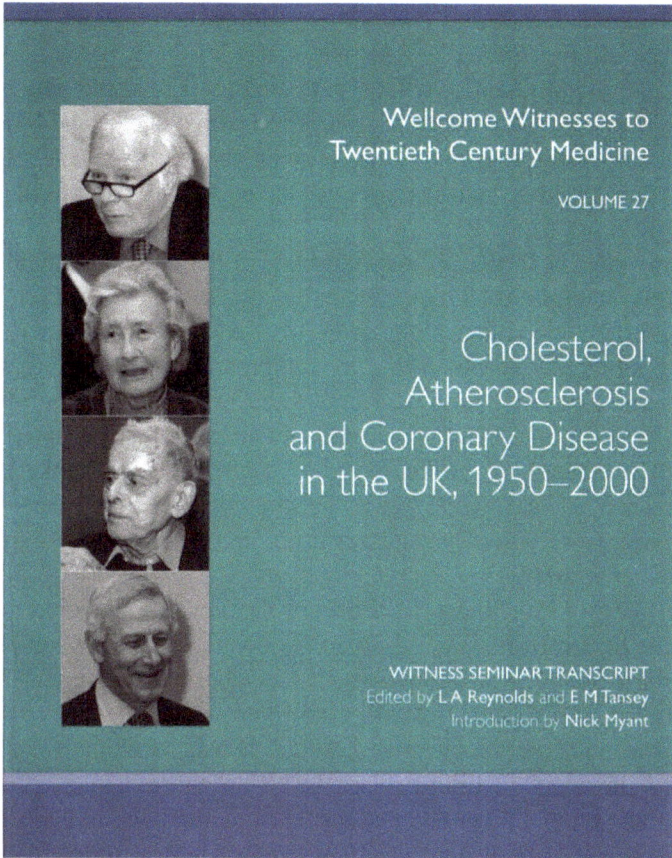

Fig. 6.1.   Front cover of Witness Seminar Transcript. Images from top: Michael Oliver, Elspeth Smith, Jerry Morris, the author.

"Vascular Sclerosis," Moschowitz from Boston stated "Arteriosclerosis cannot be prevented, no more than grey hair or wrinkles." In 1949 in the UK, Ryle and Russell wrote "It seems at present remotely unlikely that we shall discover a cure for arteriosclerosis or coronary artery disease." Enthusiasm for research into its causes or therapy did not arise before the 1950s. There was no mention of cholesterol or lipids in William Evans's influential 1948 textbook *Cardiology*. Nor of atherosclerosis as a cause of CHD.

The first mention in the UK of any relationship between cholesterol and arterial disease was in Paul Wood's First Edition (1950) of

Diseases of the Heart and Circulation. He wrote: "Lipoid substances accumulate in the intima of the large arteries in a patchy irregular fashion …. sometimes encroaching on the lumen. The lipoid nature of the deposits, their relatively frequent association with diabetes and hypercholesterolaemia, suggest some relation to fat metabolism." But, he also wrote, "consideration should be given to the possibility that raised cholesterol might result from CHD."

Meanwhile, in 1949, Gofman working in the Donner Laboratory in the Berkeley campus of UC used an analytical ultracentrifuge to show that the lipoproteins of human blood were divided into two groups — low density lipoproteins and high density lipoproteins — according to their densities and flotation rates. He labelled the low density lipoproteins as "atherosclerogenic." This concept was not immediately accepted in the USA and not at all in the UK. Oliver recalled that he spoke to Jack Gofman when he was nearly 90, who believed that his views were ignored largely because he was a physicist and not part of the orthodox medical scientific community.

In the early 1950s, three other groups had begun to develop methods to separate lipoproteins. David Barr in New York used a low temperature chemical fractionation procedure which was laborious and required 25 ml of serum. Esko Nikkilä in Helsinki used filter paper zone electrophoresis and Oliver and the biochemist George Boyd, working together in Edinburgh, developed a filter-paper zone electrophoresis micro-technique using 0.1 ml serum. Oliver recalled that their method was tedious, since it involved elution of cholesterol from 20 separate 1 cm strips of filter paper and a run of 8 hours. The latter was particularly tiresome since there was no automated switch to terminate the electrophoresis run and he and Boyd took it in turns to return to the lab between 11 pm and midnight.

In 1953, Oliver and Boyd reported the results of a study of 200 consecutive patients with coronary disease and 200 age and gender-matched controls,[2] which showed a significant elevation of plasma total cholesterol in the coronary patients at all ages. When Oliver presented these results in a ten-minute paper at the Newcastle meeting of the British Cardiac Society, not a single question was asked during the discussion period. The silence was broken eventually by the renowned Australian cardiologist Paul Wood, who dismissed Oliver and Boyd's data as "irrelevant to cardiology." However, Maurice Campbell, who was the President of the Society, commented: "Let's not be too hasty, there might just be something in this cholesterol issue."

In 1955, Oliver and Boyd reported the distribution of cholesterol between the alpha and beta lipoproteins in men aged 50 or under with coronary heart disease and age-matched controls.[3] There was 19% more cholesterol in the β fraction and an absolute decrease in the α fraction in men with CHD. The β/α ratio was 10 in the latter and 2.6 in the controls. The raised β/α ratios in men with CHD corresponded to elevated LDL/ HDL ratios in modern parlance.

Their next research venture was less successful, involving the administration of oestrogens to treat men with coronary disease. As they reported in 1961, this approach lowered both serum cholesterol and the β/α ratio, but it had feminising side effects and failed to reduce coronary events.[4] Notwithstanding this unfortunate outcome, it is fair to say that Michael Oliver's clinical access to patients with coronary disease and George Boyd's biochemical expertise with paper electrophoresis created a partnership in Edinburgh that engendered the first fruits of lipid research in Britain.

6.4. Clofibrate

As subsequently recounted by Oliver,[5] it was reported in 1954 that farm workers inadvertently exposed to an insecticide sprayed from the air over fields in France became ill and were found to have a remarkably low plasma cholesterol. The insecticide (phenyl ethyl acetic acid) had been developed by the agricultural division of Imperial Chemical Industries (ICI). An ICI chemist, Jeff Thorp, recognized the therapeutic potential of this compound and synthesized an analogue, chlorophenoxyisobutyrate.

Because of his known interest in cholesterol metabolism and plasma lipoproteins, Thorp contacted Oliver in 1957 to ask whether he and Boyd might be willing to study the cholesterol-lowering properties of this analogue. For 3 years, they explored its effects, initially in rats, and later in healthy men, starting with a dose of 250 mg daily. After a further 2 years, they showed that a daily dose of 1.5 g reduced plasma cholesterol consistently and significantly but wrongly described it as a weak androgen, androsterone. Their subsequent research showed that its action was not related to androgenic activity, but at that stage they did not understand the mechanisms through which it lowered plasma cholesterol or, to an even greater extent, plasma triglycerides.

The compound was initially designated by ICI with the trade name Atromid, and subsequently Atromid-S or clofibrate.[6] It was used in two secondary prevention trials in patients with CHD, conducted in Scotland[7] and Newcastle,[8] respectively. These trials started in 1964 and lasted for over 5 years, and both showed a significant reduction in mortality and non-fatal myocardial infarcts in patients with angina receiving clofibrate compared with those on placebo. However, there was little evidence that these effects were due to the 11–12% reduction in serum cholesterol that was observed. Other possible explanations for the reduction in CHD were the effects of clofibrate in decreasing free fatty acid and fibrinogen levels, and in reducing platelet activity.

Although the outcome of these trials must have been encouraging for the cardiologists involved, the results did little to strengthen the case for the lipid hypothesis, the only supportive piece of evidence being that patients with high baseline cholesterol levels had more events than those with lower baseline levels in the Newcastle trial.

One of the limitations of secondary prevention trials is that the outcome is determined more by the extent of myocardial damage consequent on previous infarction than by the severity of the underlying coronary atherosclerosis. On the other hand, because the incidence of myocardial infarction in middle-aged men without manifest coronary heart disease was then only about 1% per year, the sample size of a primary prevention trial would have to be much larger than a secondary prevention trial to achieve a conclusive result within 5 to 10 years.

6.5. WHO Trial of Clofibrate

With these requirements in mind, and with the support of the World Health Organization (WHO), Oliver, the epidemiologist Jerry Morris and others embarked in 1965 on a double-blind, multicentre trial of primary prevention with clofibrate in over 15,000 men aged 30–59.[9] All subjects were free from coronary heart disease at the outset and were observed for just over 5 years. The treatment group, consisting of over 5,000 men whose serum cholesterol was in the top third of the cholesterol distribution, averaging 6.8 mmol/l, received clofibrate 1.6 g/day, and a similarly chosen control group received a placebo. There was also a second control group with serum cholesterol values in the bottom third of the distribution.

Serum cholesterol decreased by 9% in the clofibrate group, and this was associated with a 25% decrease in non-fatal myocardial infarcts

($P < 0.05$) compared with the control group with comparable baseline cholesterol values. However, the incidence of fatal myocardial infarcts in the two groups was similar, and the crude death rate was actually significantly higher in clofibrate-treated subjects, mainly due to an excess of deaths from diseases involving the liver, biliary system and intestines, probably reflecting the action of clofibrate of increasing biliary cholesterol excretion. Beneficial effects of clofibrate were most obvious in individuals who smoked, who had raised blood pressure, and whose cholesterol decreased the most on the drug. The authors concluded that lowering the serum cholesterol level reduced the incidence of coronary heart disease, but that clofibrate had too many disadvantages to be used for this purpose in the general population.

The latter conclusion was reinforced 2 years later when the WHO investigators published an analysis of over 5 years of in-trial and more than 4 years of follow-up data.[10] This showed that clofibrate-treated subjects had sustained a 25% increase in total mortality, which sounded the death knell for clofibrate and posed difficult questions for adherents of the lipid hypothesis. The WHO trial undoubtedly raised serious doubts in Michael Oliver's mind regarding the safety of cholesterol-lowering therapy and influenced his attitude to reducing cholesterol by any means for many years afterwards.

6.6. Cholesterol-Lowering and Non-Cardiovascular Causes of Death

In 1985, the NIH held a Consensus Conference on Lowering Blood Cholesterol to Prevent Heart Disease.[11] After 3 days of deliberations a panel of lipoprotein experts, cardiologists, primary care physicians, epidemiologists, biomedical scientists, biostatisticians, experts in preventive medicine and lay representatives concluded that elevated blood cholesterol levels were a major cause of coronary disease and that lowering them would reduce the risk of heart attacks. It went on to recommend cholesterol screening and the treatment with diet or drugs of those found to have a raised value, as well as the adoption of dietary measures aimed at lowering the cholesterol level of the entire US population.

The findings of the Consensus Conference were criticized by Oliver[12] on the grounds that the outcome had been predetermined by the choice of experts invited to participate, although the various alternatives he proposed all seemed liable to the same source of bias. He then switched his line of attack on American efforts to prevent coronary heart disease by invoking safety concerns, pointing out that while lowering serum cholesterol reduced coronary mortality, it failed to decrease total mortality, perhaps by increasing non-cardiovascular causes of death.[13] He went on to suggest that the presence of a low cholesterol in patients with cancer was not necessarily a post-hoc phenomenon and that therapeutic lowering of serum cholesterol might sometimes cause cancer, possibly by altering the cholesterol content of cell membranes. It so happened that the expression of these safety concerns by Oliver was the prelude to a major escalation in hostilities on the cholesterol front.

6.6.1. *Meta-analysis of diet trials*

In 1990, the cardiologist Matthew Muldoon and two of his colleagues at the University of Pittsburgh, a psychologist and a psychiatrist, published a meta-analysis of six primary prevention trials of cholesterol reduction, two using diet, and four using drugs.[14] Coronary mortality was shown to be significantly reduced by drug treatment, but total mortality was unchanged. This reflected the fact that there was a significant increase in deaths from accidents, suicide and violence in those on cholesterol-lowering treatment, although there was no correlation between cholesterol reduction and deaths from cancer. At the time the authors were unable to provide a plausible explanation as to why lowering serum cholesterol should predispose to violent deaths, but they subsequently suggested that cholesterol-lowering manoeuvres might reduce the activity in the brain of serotonin, a neurotransmitter thought to exert a beneficial effect on mood, and thereby promote suicidal behaviour and aggressive tendencies.[15]

One of the studies analysed in Muldoon's original meta-analysis was the Minnesota Coronary Survey (see Chapter 5), during which inmates of six mental hospitals who had been on a lipid-lowering diet for just over a year had a significant excess of accidental deaths compared with inmates not on the diet. The causes of death included fractures, drug reactions, burns, foreign bodies, tooth extractions, freezing, heatstroke, drowning and suicide. It is difficult to conceive of a unifying causal mechanism which could explain such a disparate series of fatal events.

Evidence from other sources did not support the findings of Muldoon *et al.* In the Whitehall Study, almost 18,000 male British civil servants were followed up for 20 years, during which 90 died from accidents, violence or suicide. No significant association was seen between these deaths and serum cholesterol at entry to the study.[16] A large Swedish study did find a negative correlation between serum cholesterol at entry and suicide within the next 7 years, but not during the remainder of the 20 years of follow-up, and the increased risk of suicide was seen only in men, not in women.[17] A similar study in Finland showed no association between serum cholesterol and violent deaths or suicides during 10–15 years of follow-up but found these deaths to be more prevalent among smokers and drinkers.[18] These findings may have reassured some, but not Michael Oliver, who continued to sound the alarm in editorials with titles such as "Might treatment of hypercholesterolaemia increase non-cardiac mortality?" and "Is cholesterol reduction always safe?"

6.6.2. *Meta-analysis of diet and drug trials*

Anxieties over this issue were suddenly brought to public attention in 1992 with the publication of an article by the epidemiologists George Davey Smith and Juha Pekkanen titled "Should there be a moratorium on the use of cholesterol lowering drugs?"[19] This was at a time when the development of statins was proceeding at a rapid pace and several clinical trials were under way. Davey Smith and Pekkanen had conducted a meta-analysis similar to but larger than Muldoon's that included additional data from diet and drug trials. They confirmed that coronary mortality was reduced by lowering cholesterol in both types of trial and that total mortality was unchanged. However, they also reported a highly significant increase in non-coronary mortality, reflecting an amalgam of cancer, injury and other non-cardiovascular causes of death in the intervention groups of the drug but not in the diet trials. This led them to question whether the general use of cholesterol-lowering drugs was justified and they proposed a moratorium on the prescription of these agents until the results of the statin trials became known, still 2 years hence.

Their publication in the *British Medical Journal* resulted in an immediate plethora of hyperbolical headlines in the newspapers, ranging from "Heart pills may kill you" to "Murders linked to low-fat drugs," causing panic among patients who were on cholesterol-lowering medication and disquiet to their doctors. Public and professional anxiety regarding the safety of these compounds persisted until 1994, when the epidemiologist

Malcolm Law and his colleagues conducted an exhaustively thorough, systematic review of 40 published studies.[20] Apart from an unexplained increase in the risk of fatal haemorrhagic stroke in hypertensive subjects, they found no evidence that lowering or having low cholesterol increased mortality from any cause. They attributed the associations between low serum cholesterol and both cancer and suicide to confounding, inasmuch as low serum cholesterol is often a consequence rather than a cause of cancer or of dietary neglect, the latter being a common finding in depressed subjects with suicidal tendencies.

Oliver and his colleagues criticised Law *et al.* for conveying the impression that the excess mortality in a number of the drug trials was spurious, and they emphasised that the increased mortality of subjects on clofibrate in the WHO trial was genuine and was related to its gallstone-inducing side effect.[21] Muldoon, too, was initially reluctant to accept that cholesterol-lowering measures were safe, but he later acknowledged that "currently available evidence does not indicate that non-cholesterol illness mortality is increased significantly by cholesterol lowering treatments."[22] Despite the criticisms, Law's analysis did a great deal to reassure those responsible for prescribing lipid-lowering drugs at a time when these were under intense scrutiny and the safety of statins was still an unknown quantity.

6.7. U-Turn Over the Safety of Cholesterol-Lowering Therapy

As described in a later chapter, the publication of the results of the Scandinavian Simvastatin Survival Study (4S) in 1994 had an immediate impact on medical practice. Even Michael Oliver was impressed, and he was co-author of an article in the *British Medical Journal* titled "Lower patients' cholesterol now."[23] However, this included a caveat that the increase in non-cardiovascular mortality seen in primary prevention trials of cholesterol-lowering drugs was still an issue to be resolved. An answer was soon forthcoming, as he would have known in his role as chairman of the Data and Safety Monitoring Committee of the West of Scotland Coronary Primary Prevention Study (WOSCOPS).[24]

The results of WOSCOPS were published in 1995 and although the 22% reduction in total mortality just failed to achieve statistical significance, there was no increase in non-cardiovascular causes of death in pravastatin-treated subjects. Oliver's commentary on this occasion was titled "Statins prevent coronary heart disease,"[25] but he remained reluctant to draw any final conclusions about non-cardiac mortality, opting to wait for the results of the Heart Protection Study, still several years hence. However, he recanted before then at a meeting attended by the author that was held in The City Hall, Stockholm, in 1996 to celebrate the tenth anniversary of the award of the Nobel Prize to Goldstein and Brown. It was there that Oliver completed his U-turn, which he justified by citing Maynard Keynes: "When the facts change, I change my mind. What do you do?"

On balance, it is debatable whether Michael Oliver's considerable influence on medical opinion in his later career, especially among cardiologists, was beneficial or harmful. Based on his experiences with clofibrate, he drew attention to the need for the safety of lipid-lowering drugs. However, he also liked to speculate upon the adverse effects of lowering cholesterol per se in frequent editorials, thereby casting doubt on the treatment of patients with hypercholesterolaemia and coronary heart disease. He enjoyed creating controversy, both as a means of arriving at the truth but also because of the publicity it provided. "His heart was in the right place" is perhaps an appropriately ambivalent epitaph for this "witty, charming man with a great intellect and great (self) confidence."[26]

References

1. Oliver M. Cholesterol, atherosclerosis and coronary disease in the UK, 1950–2000. In Reynolds LA, Tansey EM (eds.), *Wellcome Witnesses to 20th Century Medicine.* 2006; **27**: pp. 4–7.

2. Oliver MF, Boyd GS. The plasma lipids in coronary artery disease. *Br. Heart J.* 1953; **15**: 387–392.

3. Oliver MF, Boyd GS. Serum lipoprotein patterns in coronary sclerosis and associated conditions. *Br. Heart J.* 1955; **17**: 299–302.

4. Oliver MF, Boyd GS. Influence of reduction of serum lipids on prognosis of coronary artery diseases — A five-year study using oestrogen. *Lancet* 1961; **ii**: 499–505.

5. Oliver M. The clofibrate saga: A retrospective commentary. *Br. J. Clin. Pharmacol.* 2012; **74**: 907–910.

6. Oliver MF. Further observations on the effects of Atromid and of ethyl chlorophenoxyisobutyrate on serum lipid levels. *J. Atheroscler. Res.* 1963; **3**: 427–444.

7. Report by a Research Committee of the Scottish Society of Physicians. Ischaemic heart disease: A secondary prevention trial using clofibrate. *Br. Med. J.* 1971; **4**: 775–784.

8. Five-year study by a group of physicians of the Newcastle upon Tyne region. Trial of clofibrate in the treatment of ischaemic heart disease. *Br. Med. J.* 1971; **4**: 767–775.

9. Report from the Committee of Principal Investigators. A co-operative trial in the primary prevention of ischaemic heart disease using clofibrate. *Br. Heart J.* 1978; **40**: 1069–1118.

10. Report of the Committee of Principal Investigators. WHO. Cooperative trial on primary prevention of ischaemic heart disease using clofibrate to lower serum cholesterol: Mortality follow up. *Lancet* 1980; **2**: 379–385.

11. Consensus Conference. Lowering blood cholesterol to prevent heart disease. *JAMA* 1985; **253**: 2080–2086.

12. Oliver MF. Consensus or nonsensus — Conferences on coronary heart disease. *Lancet* 1985; **1**: 1087–1089.

13. Oliver MF. Reducing cholesterol does not reduce mortality. *J. Am. Coll. Cardiol.* 1988; **12**: 814–817.

14. Muldoon MF, Manuck SB, Matthews KA. Lowering cholesterol concentrations and mortality: A quantitative review of primary prevention trials. *Br. Med. J.* 1990; **301**: 309–314.

15. Muldoon MF, Rossouw JE, Manuck CB, *et al.* Low or lowered cholesterol and risk of death from suicide and trauma. *Metabolism* 1993; **42**(Suppl 1): 45–56.

16. Davey Smith G, Shipley MJ, *et al.* Lowering cholesterol concentrations and mortality [Letter]. *Br. Med. J.* 1990; **301**: 552.

17. Lindberg G, Rastam L, Gullberg B, Eklund GA. Low serum cholesterol concentrations and short-term mortality from injuries in men and women. *Br. Med. J.* 1992; **305**: 277–279.

18. Vartiainen E, Puska P, Pekkanen J, *et al.* Serum cholesterol concentration and mortality from accidents, suicide, and other violent causes. *Br. Med. J.* 1994; **309**: 445–447.

19. Davey Smith G, Pekkanen J. Should there be a moratorium on the use of cholesterol lowering drugs? *Br. Med. J.* 1992; **304**: 431–434.

20. Law MR, Thompson SG, Wald NJ. Assessing possible hazards of reducing serum cholesterol. *Br. Med. J.* 1994; **308**: 373–379.

21. Heady JA, Morris JN, Oliver MF. Ischaemic heart disease and cholesterol … and mislead on adverse effects [Letter]. *Br. Med. J.* 1994; **308**: 1040.

22. Muldoon MF, Manuck SB, Mendelsohn AB, *et al.* Cholesterol reduction and non-illness mortality: Meta-analysis of randomised clinical trials. *Br. Med. J.* 2001; **322**: 11–15.

23. Oliver MF, Poole-Wilson P, Shepherd J, Tikkanen M. Lower patients' cholesterol now. *Br. Med. J.* 1995; **310**: 1280–1281.

24. Shepherd J, Cobbe SM, Ford I, *et al.* Prevention of coronary heart disease with pravastatin in men with hypercholesterolemia. West of Scotland Coronary Prevention Study Group. *N. Engl. J. Med.* 1995; **333**: 1301–1307.

25. Oliver MF. Statins prevent coronary heart disease. *Lancet* 1995; **346**: 1378–1379.

26. Watts G. Obituary: Michael Francis Oliver. *Lancet* 2015; **386**(9989): 130.

Chapter 7

Donald Fredrickson and Robert Levy: The Classification of Disorders of Lipoprotein Metabolism

7.1. Introduction

During the 1950s and 1960s inherited disorders of lipid metabolism became a subject of considerable interest to the young Donald (Don) Fredrickson at the US National Institutes of Health (NIH) in Bethesda, Maryland. He started work there in 1953 and published his first paper on lipids the following year. Life at the NIH was not all work and no play, however, and he recalled attending a symposium in 1957 in New Orleans with Ancel Keys, Alistair ("Fats") Frazer and John Youmans, Dean of Vanderbilt University. Youmans took them to a bar in the French Quarter, where they ran up a large bill buying watered-down drinks for the bar girls. He had been a jazz trumpeter in the city in his youth, so he knew the ropes and they left without paying, pursued by the bartenders and girls.[1]

Fredrickson focused his research on the role of lipoproteins in transporting lipids in plasma and on the familial disorders characterised by lipoprotein abnormalities in patients referred to the NIH. In 1961, he and his colleagues were the first to describe an inherited deficiency of HDL characterised by enlarged, orange-coloured tonsils, which they called "Tangier disease" because the index patients lived on Tangier Island in Chesapeake Bay, just off the coast of Virginia.

At the beginning of 1967, Fredrickson, Levy and Lees published a multipart paper in the *New England Journal of Medicine*, which described the phenotypic classification of five different types of hyperlipoproteinaemia.[2] Among those infected with enthusiasm for this advance were two young interns at the Massachusetts General Hospital (MGH) in Boston, Goldstein and Brown. The author was a research fellow at the MGH at that time, and he can still recall the excitement generated by Fredrickson *et al.*'s review.

Joe Goldstein subsequently looked after Fredrickson's patients as a clinical associate at the NIH, and this experience stimulated his interest in familial hypercholesterolaemia, an interest he communicated to Mike Brown. Their subsequent scientific collaboration at the University of Texas Southwestern Medical School in Dallas resulted in their discovery of the LDL receptor and the role played by genetic abnormalities of the latter in causing familial hypercholesterolaemia,[3] findings which led to them both being awarded the Nobel Prize in 1985.

Another of the interns at the MGH in 1967 was Antonio (Tony) Gotto, and he, too, came under Fredrickson's influence at the NIH. He was subsequently invited to establish a lipid research group at Baylor College of Medicine in Houston, where the author went to work with him in 1972 and 1973. The 1970s were halcyon years for clinicians and biochemists interested in lipids, but as will become apparent in later chapters, the relevance of the advances in lipidology to atherosclerosis and coronary disease was disputed by certain cardiologists, who were accused at the time of practising "paleocardiology," as distinct from "neocardiology."[4]

Fredrickson later summarised the progress which occurred between 1950 and 1975 as a Bethesda-driven "climb to base camp" that preceded the conjunction of molecular biology with the research done on lipoproteins and dyslipoproteinaemia.[5] His inspiration and leadership during that early phase of exploration were an essential preliminary to the successful scaling of scientific peaks by others, such as Goldstein and Brown. It would be hard to overestimate the influence he exerted on lipid research and research workers throughout the world during that remarkable era of progress. By differentiating between the phenotypic expression of the various types of lipoprotein disorder, Fredrickson stimulated the search for the pathological mechanisms responsible and thereby set the stage for the current era of discovery of the underlying gene defects.

7.2. The Fredrickson, Levy and Lees Classification of Lipoprotein Phenotypes

As discussed elsewhere,[6] knowledge of lipid disorders was rudimentary or non-existent among clinicians half a century ago. That changed in 1967 following the publication by Fredrickson, Levy and Lees of their five-part *magnum opus*,[2] which aimed to provide a basis upon which to manage dyslipidaemia in a rational manner. To this end, Fredrickson *et al.* introduced the idea of classifying the various disorders in terms of lipoproteins rather than lipids, substituting hyperlipoproteinaemia for hyperlipidaemia. The existence of distinct classes of lipoprotein particles differing in their lipid composition was already well established and these authors pointed out that abnormal levels of cholesterol and/or triglyceride in plasma simply reflected changes in the concentrations of the lipoproteins which transported them.

Lipoproteins can be separated according to their density on ultracentrifugation into very low density lipoprotein (VLDL), low density lipoprotein (LDL) and high density lipoprotein (HDL) or by differences in their electrophoretic mobility into the corresponding preβ, β and α lipoproteins. Because of its greater convenience, Fredrickson *et al.* adopted the latter approach as a means of determining which lipoproteins were deficient or present in excess in dyslipidaemic individuals. Drawing upon the unique collection of patients referred to the National Institutes of Health in Bethesda, they described five distinct lipoprotein phenotypes, each of which reflected genetically determined or acquired forms of hyperlipoproteinaemia.

The numbering of the phenotypes reflected the electrophoretic mobility of the lipoproteins present, chylomicrons being the slowest and α lipoprotein the fastest, with β and preβ lipoproteins exhibiting intermediate mobility.

- Type I represents an increase in chylomicrons, which remain at the origin of the strip.
- Type II is an excess of β lipoprotein (LDL).
- Type III is a unique disorder characterised by an excess of VLDL remnants with electrophoretic mobility intermediate between β and preβ lipoproteins ("broad β" band) but with the density of VLDL ("floating β") on ultracentrifugation.

- Type IV is an excess of preβ lipoprotein (VLDL).
- Type V is an amalgam of types I and IV, with an increase in both chylomicrons and preβ lipoprotein (VLDL).

Inherited forms of hypolipoproteinaemia were characterised by decreases in α lipoprotein (HDL) (Tangier disease) or β lipoprotein (LDL) (aβlipoproteinaemia and hypoβlipoproteinaemia).

The typing system was too complicated for most non-lipidologists to use in clinical practice, and as knowledge of the underlying metabolic defects increased, it gradually became redundant. However, at the time it served a vital purpose in drawing attention to the role of specific lipoproteins in predisposing to atherosclerosis, which was especially severe and premature in type II patients with an inherited increase in LDL, now known as familial hypercholesterolaemia or FH.

Fredrickson, Levy and Lees concluded their article by commenting "The plasma is often the only window from which one can see the state of intracellular metabolism. The view is limited and all ingenuity is needed to gain the sharpest perspective." The insights into lipoprotein metabolism gained from the ingenuity of these pioneers has inspired a generation of research workers and helped establish lipidology as a clinically relevant speciality.

7.3. Fredrickson: Biography

Donald Fredrickson was born in Colorado in 1924. After high school, he initially studied Medicine at the University of Colorado, but completed his studies at the University of Michigan in 1949. Between then and 1952 he worked as a resident and subsequently as a fellow in internal medicine at the Peter Bent Brigham Hospital in Boston. Subsequently he spent a year in Ivan Frantz's laboratory, which was then at the nearby Massachusetts General Hospital.

In 1953, he joined the National Heart Institute in Bethesda, Maryland, where he worked with the Nobel laureate Christian Anfinsen, and developed an interest in the metabolism of cholesterol and lipoproteins. In addition to his discovery of Tangier Disease, he played a major role in the identification and characterisation of several apolipoproteins (apoA2, apoC1, apoC2 and apoC3). From 1960 onwards he co-authored with John Stanbury and James Wyngaarden several editions of the renowned textbook *The Metabolic and Molecular Bases of Inherited Disease*. In 1978, he was awarded the Canada Gairdner International Award for his contributions to knowledge of the genetic, biochemical and clinical aspects of the hyperlipoproteinaemias.

Fredrickson was appointed director of the National Heart Institute in 1966 and then became director of the NIH between 1975 and 1981 (Fig. 7.1). One of the main issues that occupied him there was the controversy over research involving recombinant DNA. He drew up a guideline that restricted release of genetically modified organisms into the environment and established a committee that had to approve any NIH research involving recombinant DNA technology, thus restoring confidence in this form of research. In 1983, he became the vice-president of the Howard Hughes Medical Institute but returned to the NIH in 1987 to resume lipid research.

In 2002, Don Fredrickson was sadly found dead in his swimming pool and is buried in Leiden in his wife's homeland, Holland. In his obituary, Tony Gotto described him as an absolutely honest, trustworthy and candid individual who, as a scientist, clinician, administrator and public servant

Fig. 7.1.   Don Fredrickson (1924–2002) (Photo: NIH).

adhered to the highest standards of performance and integrity and was a towering influence in the scientific community.[7]

7.4. The Lipid Research Clinics Coronary Primary Prevention Trial

The Lipid Research Clinics (LRC) Program had been established in 1970 by the National Heart and Lung Institute and was headed initially by Bob Levy and later by Basil Rifkind, a Glaswegian cardiologist who had recently emigrated to the USA. One of its major research endeavours was the Lipid Research Clinics Coronary Primary Prevention Trial (LRC–CPPT), which took place in 12 centres during the 1970s and was completed during Levy's tenure as director of the National Heart, Lung and Blood Institute (previously National Heart Institute) in succession to Fredrickson.

The LRC-CPPT was a multicentre, double-blind trial of cholesterol lowering in the primary prevention of coronary heart disease in North America.[8] This involved screening almost 500,000 men aged 38–59, of whom 3,806 had a serum total cholesterol level higher than 6.8 mmol/l and were found to be free from overt coronary heart disease. After randomisation, the test and control groups were placed on a cholesterol-lowering diet; in addition, the test group was given 24 g cholestyramine per day, an unabsorbed anion exchange resin that lowers cholesterol by binding bile acids in the intestine and inhibiting their reabsorption, thereby promoting uptake and conversion of LDL cholesterol into bile acids by the liver to compensate for their loss, whereas the control subjects received a matching placebo.

After 7 to 10 years, average levels of serum total and LDL cholesterol in the cholestyramine group were 8.5% and 12.6% lower, respectively, than those of the placebo group. There was a 19% reduction in deaths from coronary heart disease plus non-fatal myocardial infarcts in the cholestyramine group, but no significant difference between the two groups in total mortality.

To ascertain whether the reduction of risk was related to the degree of reduction of hypercholesterolaemia, the 155 men in the cholestyramine-treated group who had sustained a coronary event during the trial were subdivided according to the year when this occurred.[9] The reduction of LDL cholesterol in these individuals was then compared with the reduction in LDL in those who had remained free from coronary heart disease over the same period. The results suggested that decreasing serum total cholesterol by 26% or LDL cholesterol by 35% would be expected to halve the risk of developing coronary heart disease in hypercholester-olaemic men.

This all happened at a time when the entire future of lipid-lowering therapy was in doubt as a result of the negative publicity generated by earlier trials such as the WHO trial of clofibrate. Hearsay has it that when the results of the LRC–CPPT were being analysed, the statisticians involved were locked in a room at the NIH and told they would be let out only after they had come to a favourable decision. Eventually, a note came out under the door asking whether a reduction in coronary events which was significant at the 5% level on the basis of a one-tailed *t*-test was acceptable! It was for the authors but not for some statisticians and cardiologists like Michael Oliver, who insisted on the more stringent two-tailed *t* test as the arbiter of statistical significance. Tony Gotto happened to be president of the American Heart Association (AHA) at the time and he doubts whether the AHA would have endorsed the trial outcome had he not been.[10]

Although there was no decrease in total mortality, the results of the LRC-CPPT were sufficiently promising to convince many of the doubters that cholesterol-lowering therapy was effective in preventing coronary disease and thereby ensured its future as a clinical endeavour. For that reason, and many others besides, Bob Levy earned his place in history alongside his boss and predecessor at the NIH, Don Fredrickson. Between them they provided the basis of knowledge upon which the diagnosis and management of dyslipidaemia currently rests. The remarkable decrease in coronary heart disease in the USA since 1980, almost a quarter of which has been attributed to reductions in serum cholesterol,[11] is due in no small part to their conjoint efforts.

7.5. The NHLBI Type II Coronary Intervention Study

Published almost simultaneously with the LRC Primary Prevention Trial in 1984, this double-blind study was the first of a series of angiographic trials performed between 1984 and 1994. In this study, patients with type II hyperlipoproteinaemia (i.e. with raised LDL cholesterol) were randomised to receive a cholesterol-lowering diet plus cholestyramine 24 g daily or diet plus placebo. LDL cholesterol decreased by 5% in those on placebo and by 26% in those on cholestyramine. Coronary angiography was performed in 116 patients before the start of the trial and again 5 years later. This showed that coronary disease had progressed in 49% of placebo-treated patients versus 32% of those on cholestyramine, the difference being statistically significant. The sample size was too small to draw any definitive conclusions, but the results suggested that lowering LDL probably retarded the progression of coronary artery disease in patients with a raised LDL cholesterol.[12] In this trial and in several others,[13] the response to treatment was greatest in coronary artery lesions that were >50% stenosed at baseline.

As in other aspects of lipid research in the late 1970s and early 1980s, the NHLBI under Bob Levy's leadership led the way in angiographic studies of the effects of cholesterol-lowering on established atherosclerosis. Later studies were able to use statins and thereby achieve much greater reductions in LDL than with cholestyramine and hence obtain more definitive evidence of benefit.

7.6. Levy: Biography

Robert (Bob) Levy was born in the Bronx, New York, in 1937. After graduating from the Bronx High School of Science, he entered Cornell University in Ithaca at the age of 16. After a highly successful career as a student at Cornell, he went on to Yale University Medical School and did internships at Yale New Haven Hospital. He joined the lipid metabolism branch of the National Heart, Lung, and Blood Institute (NHLBI) in 1963 and succeeded Fredrickson as Director in 1975. He left the NHLBI in 1981 to take up administrative posts at Tufts University School of Medicine and Columbia University College of Physicians and Surgeons. He went on to become President of the Sandoz Research Institute from 1988 to 1992 and then joined American Home Products as President of its Wyeth-Ayerst Research Division in 1992. In 1998, Bob Levy became the Senior Vice President for Science and Technology at American Home Products.

From trainees to patients, to staff, to peers, he had the ability to communicate a genuine interest in people and in their activities — for example, he was a passionate supporter of the Washington Redskins football

Fig. 7.2.   Bob Levy (1937–2000) with the author at Stonehenge.

team. At a personal level he and Don Fredrickson were as different as chalk from cheese — Levy was brought up in the Bronx and his direct manner antagonised some people, whereas Fredrickson's urbane sophistication could charm a bird off a tree. Nevertheless, Bob's ability to form lasting friendships with individuals can only be explained by his having a genuine interest in their welfare and the author greatly valued the friendship and good advice he received from Bob regarding the future direction of his own research (Fig. 7.2). Bob Levy's strong personality combined with his imaginative research and leadership qualities made him a truly unique and unforgettable individual.[14] He died in New York in 2000.

7.7. Role of the NIH in Establishing the Lipid Hypothesis

In the latter half of the 20th century the NIH, in the shape of the NHLBI, was the leading centre of research into lipids, specifically of cholesterol-rich and triglyceride-rich lipoproteins, and their role in atherosclerosis and cardiovascular disease. For clinicians working in lipid clinics the classification of lipoprotein phenotypes provided a compass to the nature of the abnormalities in their patients and hence clues as to the possible cause and therefore the most profitable line of treatment. For example, patients with a type II phenotype often had a family history suggestive of a genetic basis and their raised levels of LDL cholesterol responded best to cholest-yramine; there were no statins in the 1970s and early 1980s. In contrast, hypertriglyceridaemia due to a raised VLDL in patients with a type IV phenotype was often due to type II diabetes and responded best to a fibrate, as did increased levels of VLDL remnants in type III patients.

Clinical research in those days often involved in vivo studies of the turnover of lipoproteins labelled with radio-isotopic cholesterol or apoli-poprotein B. For example, in 1972 Levy and colleagues showed that the fractional rate of turnover of ^{125}I-labelled LDL was reduced in FH patients,[15] implying the existence of an in vivo catabolic defect. The underlying deficiency of LDL receptors responsible was revealed the following year by Goldstein and Brown's in vitro studies.[3]

Studies with cultured fibroblasts took off in the aftermath of Goldstein and Brown's research on the binding and degradation of labelled LDL in vitro. More recently, DNA analysis has become part and parcel of the diagnostic armoury of lipid clinics and has transformed our understanding of the basis of many forms of hyper- and hypo-lipidaemia, enabling clinicians to match phenotype (clinical and biochemical charac-teristics) with genotype (DNA profile). It is impossible to know whether or how soon these advances would have come about without the pioneer-ing research conducted at the NIH by Fredrickson and Levy and their colleagues.

References

1. http://profiles.nlm.nih.gov/FF/B/B/L/G/_/ffbblg.txt.
2. Fredrickson DS, Levy RI, Lees RS. Fat transport in lipoproteins — An inte-grated approach to mechanisms and disorders. *N. Engl. J. Med.* 1967; **276**: 34–42, 94–103, 148–156, 215–225, 273–281.
3. Goldstein JL, Brown MS. Familial hypercholesterolemia: Identification of a defect in the regulation of 3-hydroxy-3-methylglutaryl coenzyme A reduc-tase activity associated with overproduction of cholesterol. *Proc. Natl. Acad. Sci. USA* 1973; **70**: 2804–2808.
4. Carruthers M, Taggart P. Paleocardiology and neocardiology. *Am. Heart J.* 1974; **88**: 1–6.
5. Fredrickson DS. Phenotyping. On reaching base camp (1950–1975). *Circulation* 1993; **87**(Suppl 4): III 1–15.

6. Thompson GR. Lipids, dyslipidaemia and cardiovascular disease. In: *Understanding Medical Research*, Goodfellow JA (ed.). Oxford: Wiley-Blackwell; 2012.

7. Gotto AM, Jr. Donald S. Fredrickson, MD 1924–2002. *Arterioscler. Thromb. Vasc. Biol.* 2002; **22**: 1506–1508.

8. Lipid Research Clinics Program. The lipid research clinics coronary primary prevention trial results. I. Reduction in incidence of coronary heart disease. *JAMA* 1984; **251**: 351–364.

9. Lipid Research Clinics Program. The lipid research clinics coronary primary prevention trial results. II. The relationship of reduction in incidence of coronary heart disease to cholesterol lowering. *JAMA* 1984; **251**: 365–374.

10. Gotto AM, Olsson AG. *The Cholesterol Story*. Chisinau, Republic of Moldova Europe: LAP LAMBERT Academic Publishing; 2021.

11. Ford ES, Ajani UA, Croft JB, *et al.* Explaining the decrease in US deaths from coronary disease, 1980–2000. *N. Engl. J. Med.* 2007; **356**: 2388–2398.

12. Brensike JF, Levy RI, Kelsey SF, *et al.* Effects of therapy with cholestyramine on progression of coronary arteriosclerosis: Results of the NHLBI type II coronary intervention study. *Circulation* 1984; **69**: 313–324.

13. Thompson GR. Angiographic trials of lipid-lowering therapy: End of an era? *Br. Heart J.* 1995; **74**: 343–347.

14. Gotto AM, Jr. In Memoriam: Robert I. Levy 1937–2000. *J. Lipid Res.* 2001; **42**: 886.

15. Langer T, Strober W, Levy RI. The metabolism of low density lipoprotein in familial type II hyperlipoproteinemia. *J. Clin. Invest.* 1972; **51**: 1528–1536.

Chapter 8

John Yudkin: The Role of Sugar in Coronary Heart Disease

8.1. Introduction

The early 1970s were a period of transition for those working in the lipid field, mainly spent digesting the advances in lipoprotein phenotyping introduced by Fredrickson and his colleagues at the NIH and trying to keep pace with the ever-lengthening alphabet of recently discovered apolipoproteins. In contrast to the rapid advances in knowledge of the structure and function of lipoproteins, early attempts to influence the progression of atherosclerosis and prevent coronary disease by lowering serum cholesterol by diet or drugs were relatively unrewarding. The absence of any definitive evidence that lowering cholesterol influenced clinical outcome rendered the lipid hypothesis open to doubt and left room for alternative explanations for the rise in coronary mortality since the end of the Second World War. This trend reached its peak in 1970 when the UK was second only to Finland in the world rankings.

8.2. Sugar and Coronary Heart Disease

In the summer of 1974, *The Times* of London published an article titled "Why suspicion falls on sugar as a major cause of heart disease." The author was John Yudkin, Emeritus Professor of Nutrition at Queen Elizabeth College, University of London, whose work at the time was supported by the National Dairy Council.[1] Yudkin had previously reported that the increase in coronary disease in Britain was proportional to the increase in radio and television licences and he suggested that the evidence linking saturated fat, serum cholesterol and coronary disease was in a similar category, lacking proof of causality. In particular, the lack of evidence from dietary trials that modifying fat intake reduces the risk of coronary heart disease made him conclude that the lipid hypothesis was defunct. Among the alternative explanations he proposed were a lack of roughage or trace elements in the diet, and an excess of protein, carbohydrate or sugar.[2]

Consumption of sugar in Britain had increased 25-fold during the past 200 years, and Yudkin cited data that a high intake of sucrose could raise serum cholesterol and triglyceride, especially the latter, as well as levels of insulin, cortisol and uric acid, and could also impair glucose tolerance and induce a clotting tendency. All these abnormalities were evident in subjects with coronary disease, so he claimed, and he proposed that they reflected a hormonal disturbance brought about by excessive consumption of sucrose. He concluded that eating less sugar would correct the hormonal imbalance and also favourably influence dental disease, diabetes and obesity, whereas altering dietary fat intake, particularly increasing the intake of polyunsaturated fats, would achieve none of these objectives and might have adverse side effects.

Earlier that year Yudkin had contributed to a Report of the Committee on Medical Aspects of Food Policy (COMA) on diet in relation to cardiovascular disease in the UK.[3] The majority of the other members of the committee recommended that the amount of total and saturated fat in the national diet should be reduced to decrease cardiovascular disease and that sucrose consumption should be reduced to decrease obesity. However, in a note of reservation at the end of the report, Yudkin recorded his opinion that the role of dietary fat in causing cardiovascular disease had been exaggerated and that the role of sucrose had been minimised. One of the other members of COMA, JRA Mitchell, was subsequently highly critical of Yudkin, especially his citing as evidence unproven speculations bearing no relationship to the matter under discussion.[4]

8.3. Paucity of Evidence

Yudkin's claims regarding the link between sugar and vascular disease were based on remarkably scanty experimental evidence. In 1964, he and

Roddy[5] had shown that the estimated intake of sucrose in 45 men with coronary or peripheral vascular disease was significantly greater than that of 25 control subjects. On the basis of these results, the authors claimed that the relationship between sucrose and vascular disease was causal and that people taking over 110 g/day were five times more likely to develop myocardial infarction than those taking under 60 g/day. Three years later, following criticisms of the methodology used to estimate sucrose intake, Yudkin and Morland carried out another study, modified to minimize any bias in estimating dietary intake, in 20 myocardial infarct survivors and 33 controls, and again found a significantly higher intake of sucrose among the former.[6]

In the light of these findings, the Medical Research Council (MRC) convened a working party to investigate the relationship between sugar and vascular disease. Studies were undertaken in four centres, two in London and two in Scotland. The largest study took place at the Central Middlesex Hospital, where the epidemiologist Richard Doll, famous for his work on the dangers of smoking, was involved while the study conducted at the Edinburgh Royal Infirmary was supervised by Michael Oliver. The pooled results of all four studies showed no difference between the sucrose intake of 122 myocardial infarct survivors and 113 controls, leading the working party to conclude, in 1970, that the evidence in favour of a high sugar intake as a major factor in the development of myocardial infarction was extremely slender.[7]

The following year Ancel Keys published a scathing review of Yudkin's work stating that there was no theoretical basis or experimental evidence to support his claim for a major influence of dietary sucrose in the aetiology of CHD. Keys concluded that any apparent association of sucrose with CHD was in fact due to close correlations between sugar consumption, saturated fat intake and cigarette smoking. He considered that the latter two factors were causally related to CHD whereas evidence of causality for sucrose was not supported by any clinical, epidemiological or experimental evidence.[8]

8.4. Literary and TV Debates

Despite the MRC working party's and Key's criticisms, Yudkin persevered in his beliefs, which he had not only publicised in a book[9] but, as noted earlier, he also reiterated 4 years later in *The Times*. The author persuaded that newspaper to allow him to rebut Yudkin's claims and put the case for saturated fat, rather than sucrose, being the dietary cause of cardiovascular disease. The article was titled "Beware sweet reason in the search for causes of heart disease,"[10] and it concluded that the evidence linking sucrose and cardiovascular disease was largely circumstantial, similar to that linking TV sets and heart attacks, whereas the evidence that saturated fat was harmful was far stronger.

The Times article caught the eye of a television producer at the BBC, who arranged for a debate to take place at The Royal Institution in September 1974, which was recorded and transmitted subsequently in the *Controversy* series on TV. The motion to be debated was "The dietary cause of heart disease is sugar, not fat," proposed by John Yudkin and opposed by Don Fredrickson, assisted by Peter Taggart, a cardiologist at the Middlesex Hospital and the author, with Sir George Porter, Nobel laureate in chemistry, in the chair. Despite being heavily outnumbered, Yudkin put up a stiff resistance and kept bouncing back like a punch ball despite repeated verbal knockdowns.

After the debate was over, Don Fredrickson and the author had dinner together in a nearby restaurant. At that stage Fredrickson was president of the Institute of Medicine of the National Academy of Sciences in Washington and about to become director of the NIH, although one would never have guessed it from his unassuming manner. In that role, he soon became involved in the controversy over recombinant DNA research and was instrumental in getting this legitimised in the face of considerable opposition. He was a man of great charm and wit, an exception to the rule that "nice guys finish last!"

8.5. Sugar and the Metabolic Syndrome

The sucrose hypothesis initially faded away, although the biochemical abnormalities which Yudkin attributed to excess sugar consumption have since been recognised as part of the metabolic syndrome. The latter is an accompaniment of the central obesity, which is becoming increasingly common nowadays as a result of excessive caloric intake, especially when derived from rapidly absorbed carbohydrates such as sucrose. Individuals with the metabolic syndrome have an increased liability to develop hypertriglyceridaemia and coronary heart disease, so in this respect Yudkin was right, as was acknowledged recently.[11] Where he erred was in overemphasising the role of sucrose as a risk factor for CHD, despite the lack of evidence for this, and in disregarding the role of saturated fat, despite the evidence in its favour. Official acceptance of his views could have had unforeseen consequences for the population approach to prevention of coronary heart disease in Britain, but, happily, this did not transpire. The only reference to sugar in the subsequent recommendations from COMA 10 years later regarding cardiovascular prevention was that the intake of sucrose, glucose and fructose should not increase any further.[12]

8.6. Commercial Influences

Yudkin's views on diet and heart disease must have been music to the ears of the National Dairy Council. The late Keith Ball, one of the earliest advocates of dietary change to prevent coronary disease, criticised the massive advertising campaign conducted by the Butter Information Council and the National Dairy Council to persuade doctors and the public to disregard official advice to reduce saturated fat intake.[13] He claimed that general practitioners had been circulated with the names of experts who disagreed with the views expressed by official bodies such as COMA, and he questioned the ethics of such promotional activities. Yudkin himself, of course, had openly dissociated himself from COMA's views on fat intake while he was a member of the committee in 1974.[3]

Yudkin's best-seller *Pure, White and Deadly* was originally published in 1972. A revised version was published in 1984 and re-published in 2012. The book begins by tracing the history of cane sugar as a profitable product of the slave trade in the Caribbean and USA. Up to now sugar seems to have escaped censure by the Black Lives Matter movement. However, the Tate Galleries' website currently emphasises that although Sir Henry Tate (the sugar refiner) was not a slave-owner or slave-trader, it is impossible to disassociate the Tate Galleries in the UK from the colonial slavery from which their benefactor's wealth was derived historically.

Yudkin's book goes on to describe experiments he conducted in young men showing that a sugar-rich diet raised their serum triglycerides and in a minority of instances increased their serum insulin levels. He asserted that diabetes was four times more prevalent in Asians than in Europeans and cited evidence that the former consumed more sugar than the latter, implying that this may be the reason why being Asian is a risk factor for coronary heart disease.

At the start of his book Yudkin states "I hope that when you have read this book I shall have convinced you that sugar is really dangerous." This message was extremely unwelcome to the sugar industry and he listed several examples of attempts to interfere with the funding of his research and prevent its publication. In 1984, he won a libel action with costs against the World Sugar Research Foundation and forced it to publish an apology for commenting in its quarterly bulletin in 1979 that "Readers of science fiction will no doubt be distressed to learn ... that *Pure, White and Deadly* is out of print"! It may have been out of print then, but now, more than 40 years later, it is widely available.[14]

In his final chapter, Yudkin acknowledged that "by no means every scientist agrees with my views about sugar. And there is nothing wrong with that: much of the material which I have written still adds up to circumstantial evidence rather than absolute proof ... that sugar is, for example, one of the causes of coronary disease." Thus, he hedged his bets in later life, in contrast with his more aggressive stance in 1974, when he accused the author of libel for stating during the BBC TV debate that the evidence that sugar was atherogenic was largely circumstantial!

Commerce and academia sometimes make uncomfortable bedfellows, as illustrated by the 2003 recommendation by the World Health Organisation (WHO) that sugar consumption should be limited to 10% of energy intake, a proposition rejected by the food industry but promulgated by the WHO. Reducing the amount of sugar consumed in soft drinks seems to be especially important because of the ease with which sugar is ingested in this form[15] and the role it consequently plays in childhood obesity, as Yudkin stressed in his book (Fig. 8.1).

8.7. Biography

John Yudkin was born in London in 1910, the son of Jewish immigrants who, like John Gofman's parents, had fled from Russia to escape the pogroms. After schooling in London he won a scholarship to Cambridge University and obtained a BSc in Biochemistry in 1931, followed by a PhD in 1935. While working for his PhD he taught biochemistry and physiology to medical students at Cambridge. He then started studying Medicine himself and qualified at the London Hospital in 1938. He began

Fig. 8.1.   John Yudkin (1910–1995).

doing research on dietary vitamins at the Dunn Nutritional Laboratory in Cambridge, but the Second World War intervened, during which he served as a medical officer in the Royal Army Medical Corps in Sierra Leone. After the war he was appointed to the Chair in Physiology at Queen Elizabeth College, London, where he set up a BSc degree course in Nutrition. In 1954, he was made Professor in charge of the newly established Department of Nutrition at Queen Elizabeth College. He retired as head of the department in 1971 and ceased working at Queen Elizabeth College in 1974. He died in London in 1995, aged 85.

John Yudkin will be remembered not just for his controversial views on sucrose but also as the holder of the first Chair of Nutrition in the UK. He was highly regarded in nutrition circles and were he alive today, he would have been pleased that the latest American College of Cardiology/American Heart Association guideline on the primary prevention of cardiovascular disease recommends minimising the dietary intake of refined carbohydrates and sweetened beverages.[16]

References

1. Yudkin J. Why suspicion falls on sugar as a major cause of heart disease. *The Times* 1974; 11 July, p. 16.

2. Yudkin J. Sucrose and cardiovascular disease. *Proc. Nutr. Soc.* 1972; **31**: 331–337.
3. Department of Health and Social Security. Diet and coronary heart disease: Report of the Advisory Panel of the Committee on Medical Aspects of Food Policy (Nutrition) on diet in relation to cardiovascular and cerebrovascular disease. London: HMSO 1974 (Report on Health and Social Subjects, no. 7).
4. Mitchell JRA. Diet and coronary heart disease — A British point of view. *Adv. Exp. Med. Biol.* 1977; **82**: 823–827.
5. Yudkin J, Roddy J. Levels of dietary sucrose in patients with occlusive atherosclerotic disease. *Lancet* 1964; **41**: 6–8.
6. 6.Yudkin J, Morland J. Sugar intake and myocardial infarction. *Am. J. Clin. Nutr.* 1967; **20**: 503–506.
7. Dietary sugar intake in men with myocardial infarction. Report to the Medical Research Council by its working party on the relationship between dietary sugar intake and arterial disease. *Lancet* 1970; **2**: 1265–1271.
8. Keys A. Sucrose in the diet and coronary heart disease. *Atherosclerosis* 1971; **14**: 193–202.
9. Yudkin J. *Pure, White and Deadly: The Problem of Sugar*. London: Davis-Poynter; 1972.
10. Thompson GR. Beware sweet reason in the search for causes of heart disease. *The Times* 1974; 31 July, p 16.
11. Watts G. Sugar and the heart: Old ideas revisited. *BMJ* 2013; **346**: e7800. doi: 10.1136/bmj.e7800.
12. Department of Health and Social Security. Diet and Cardiovascular Disease: Report of the Panel on Diet in Relation to Cardiovascular Disease, Committee on Medical Aspects of Food Policy. London: HMSO, 1984 (Report on Health and Social Subjects, no. 28).
13. Ball K. Prevention of coronary heart-disease. *Lancet* 1979; **2**: 1182.
14. Yudkin J. *Pure, White and Deadly: How Sugar is Killing Us and What We Can Do to Stop It.* London: Penguin Books; 2012.
15. Willett WC, Lustig DS. Science souring on sugar. *BMJ* 2013; **346**: e8077. doi: 10.1136/bmj. e8077.
16. Arnett DK, Blumenthal RS, Albert MA, *et al.* ACC/AHA guideline on the primary prevention of cardiovascular disease: Executive summary: A report of the American College of Cardiology/American Heart Association task force on clinical practice guidelines. *J. Am. Coll. Cardiol.* 2019; **74**: 1376–1414.

Chapter 9

Sir John McMichael and J.R. Anthony Mitchell: Cardiological Opponents of the Lipid Hypothesis

9.1. Leader of the Opposition

Sir John McMichael was once described as the greatest clinical scientist of his generation.[1] He directed the Department of Medicine at the Royal Postgraduate School at the Hammersmith Hospital, London, from 1946 to 1966, and to those who worked under him, including the author, he was an Olympian figure akin to "M" in the James Bond films, evoking in equal measure respect, admiration and affection. McMichael's ability to create an atmosphere of enthusiasm, constructive criticism and cooperation was such that the best young research workers were attracted by the reputations of the people working there, and this earned worldwide recognition for the Hammersmith as the best place in the British Commonwealth for clinical research and postgraduate medical training.[2]

Following his retirement from the Hammersmith in 1966, McMichael became director of the British Postgraduate Medical Federation and continued to influence postgraduate medical training for several years. In 1975, he gave the Harveian Oration at the Royal College of Physicians, titled "A transition in cardiology: the Mackenzie–Lewis era."[3] During his discourse, he commented, "Resistance to change is far from unknown among great scientists," and he exemplified this by Virchow's rejection of Koch's pioneering work. He went on to assert that the cardiologists Mackenzie and Lewis's "obstinate Celtic temperaments caused them to defend their view" despite overwhelming evidence to the contrary. Ironically, both these statements characterised his own attitude to the lipid hypothesis.

9.2. Letters to *The Lancet*

In 1973, 3 years after he retired from the British Postgraduate Medical Federation and long after his departure from the Hammersmith, McMichael commenced hostilities with those who believed that diet and exercise played a role in the aetiology of coronary heart disease. In the first of several letters to *The Lancet* he criticised the International Society of Cardiology for stating that an increased intake of cholesterol and fat is a prerequisite for the development of atherosclerosis and took issue with Jerry Morris's data from bus drivers and conductors of double-decker buses showing that exercise had a protective effect.[4]

Two years later, McMichael returned to the fray to criticise the recommendations of a joint working party of the Royal College of Physicians and the British Cardiac Society on the Prevention of Coronary Disease, which endorsed the policy of reducing saturated fat intake to lower serum cholesterol.[5] He reiterated his conviction that this measure would have little effect on the incidence of coronary disease and concluded that "it is better to trust to luck than to foster neurosis by pretence that we can save lives by interfering with life habits." The chairman of the joint working party, the South African-born epidemiologist Gerry Shaper, responded that the measures recommended in the report had a reasonable hope of conferring benefit, and none had a cost that approached the cost of inaction.[6] McMichael was unimpressed and again criticised the report for its simplistic dietetic advice and the manner in which its views were publicised by the Department of Health, which had sent a copy to every doctor.[7] Further support for the report came from the Hammersmith cardiologist John Goodwin, who had been a member of McMichael's department from its early days. He had instigated the joint working party when he was president of the British Cardiac Society, and he defended the advice it gave to the medical profession aimed at preventing coronary disease, while accepting that knowledge on this topic was incomplete.[8]

McMichael next turned his attention to the recently published results of the Whitehall Study of civil servants by Geoffrey Rose and his colleagues and questioned whether the correlation of raised cholesterol with

Fig. 9.1.   Portrait of Sir John McMichael (1904–1993) by Sir William Hutchison.

coronary risk in patients with angina or an abnormal ECG might be a consequence of myocardial ischaemia rather than a cause, as his Hammersmith colleague Paul Wood had suggested many years previously. He also queried these authors' advocacy of the need for further prevention trials.[9] Rose, an outstanding epidemiologist, and his colleagues politely pointed out that McMichael's conclusions were based on a misinterpretation of their findings,[10] Many of McMichael's criticisms pertaining to lipids and epidemiology were scientifically naive, but his past achievements and considerable reputation ensured his arguments were well publicised (Fig. 9.1).

9.3.   European Society for Clinical Investigation Debate

The chance to put his views to the test came in 1977 when the European Society for Clinical Investigation held a "Controversy in medicine" debate at its annual meeting in Rotterdam. The motion to be debated was

"That modification of serum lipids by dietary and/or other means will influence the incidence of, or mortality from, coronary heart disease." In favour of the motion were Lars Carlson from Stockholm, Shlomo Eisenberg from Jerusalem, and the author from London. Carlson was the premier clinical lipidologist in Europe at the time while Eisenberg had established a major reputation as a result of the research he had carried out with Levy at the NIH. Against the motion were Sir John McMichael and Professors Paul Astrup and Christian Crone, both from Copenhagen. The debate was chaired by Hermon Dowling, who was then the Secretary of the European Society.

The proponents of the motion put forward a tripartite case for there being a causal association between hyperlipidaemia and coronary disease and postulated that reversal of the former would reduce the incidence of the latter. This was based on three lines of scientific evidence, epidemiological, experimental and clinical, presented sequentially. The case against the motion was put by Christian Crone, who simply showed two slides. The first was a photograph of a sleek, well-groomed rat, which, he said, had been fed on butter from birth, while the second showed a malnourished rodent with dull, moulting fur, which, he claimed, had been fed on corn oil throughout its life. The audience dissolved into laughter, and Crone's debating ploy ensured that the motion was lost! McMichael was delighted at the outcome, but Carlson was furious.

9.4. Fats and Atheroma: An Inquest

As mentioned earlier, 1978 saw the publication of the results of the WHO trial of clofibrate, which showed an increase in non-cardiovascular causes of death in those on the drug. McMichael was not slow to capitalise on this finding, and in January 1979 he published a paper in the *British Medical Journal* titled "Fats and atheroma: an inquest."[11] In this, he alluded to the negative results of the early diet trials as casting doubt on the role of cholesterol in atherogenesis, and he pointed out the potential dangers of lowering blood levels with polyunsaturated fats and clofibrate, namely cancer of the colon and gallstones, respectively. He concluded that official medical endorsement of these cholesterol-reducing measures should be withdrawn. In a similar paper published later that year, he characterised the campaign to substitute polyunsaturated for saturated fats as a "quite unwarranted extension of hope over experience."[12]

9.5. Lobbying the Royal College of Physicians

In 1979, JRA Mitchell and McMichael, both of whom were Fellows of the Royal College of Physicians, published a letter together in the *Journal of the Royal College of Physicians*, suggesting that the college should with-draw its support for the dietetic recommendations put forward 3 years previously by its joint working party with the British Cardiac Society.[13] McMichael raised the matter at the Comitia (annual general meeting) of the college, and it was then referred to the Council of the college, which, under the enlightened presidency of Sir Douglas Black, declined to reopen the issue. McMichael was not to be silenced, however, and raised the issue again several times in 1980. Eventually, the college agreed to undertake a further report on the prevention of coronary heart disease, but only after the benefits and risks of polyunsaturated fats had been considered by a WHO-sponsored International Meeting of Cardiologists in 1982. This decision did not placate McMichael, who expressed his displeasure at Comitia in January that year.

9.6. Farewell Staff Round

In 1985, a special staff (grand) round was held at the Hammersmith to commemorate Sir John McMichael's 80th birthday. Sadly, he had by then sustained a severe stroke and was confined to a wheelchair. The author presented one of the cases, a patient with severe hyperlipidaemia who had responded well to lipid-lowering therapy. Sir John nodded and smiled, but his aphasia made it impossible to know whether he had changed his mind about cholesterol. It seemed improbable but not impos-sible, for in his Harveian Oration he had quoted a passage from William Harvey's *De Motu Cordis*: "Studious good and honest men do not think it degrading to alter their view if truth and a public demonstration so persuade them."

9.7. McMichael: Biography

John McMichael was born in 1904 in Gatehouse of Fleet in Scotland. After local schooling he studied Medicine at Edinburgh and qualified in 1927. From there he went south to work as a Beit Memorial Fellow at University College Hospital, London, and obtained an MD. In 1938 he was appointed Reader in Medicine at the newly established Postgraduate Medical School at the Hammersmith and became Professor of Medicine there in 1946.

In 1960, he was awarded the Canada Gairdner International Award in recognition of his contributions to cardiology and clinical physiology and especially for his achievements in the early application of cardiac catheterisation for the measurement of cardiac output, thus making an important contribution to the understanding of heart failure. He had numerous honorary degrees and accolades bestowed upon him during his distinguished career including Fellowship of the Royal Society and a knighthood. He retired from the Hammersmith in 1966 and died in 1993, aged 89.

After he died, a service of thanksgiving was held for him at St Columba's Church of Scotland in London. It was packed with the Hammersmith staff, past and present, united in paying tribute to a great man. As his successor Chris Booth said in his obituary, McMichael had created a clinical research environment, unique in Britain at that time, of free discussion and debate in which all, whatever their level of seniority, might join.[1]

9.8. Mitchell: "The Abominable No Man"

One of the keenest minds and sharpest tongues among critics of the lipid hypothesis belonged to JRA (Tony) Mitchell, the foundation Professor of Medicine at Nottingham University from 1968 until his retirement in 1990. His early research was carried out in the department of the Regius Professor of Medicine at Oxford, where he and Colin Schwartz studied the pathology of atherosclerosis. In the book they published on the subject, they proposed that turbulent blood flow at sites of arterial branching caused platelets to coalesce and form mural thrombi, resulting in the formation of raised plaques on the arterial intima.[14] This concept was in keeping with the Duguid or thrombogenic theory of atherogenesis but was contrary to the lipid hypothesis, the validity of which Mitchell constantly questioned.

9.9. COMA

Tony Mitchell was a member of the first Committee on Medical Aspects (COMA) of diet and coronary heart disease, which published its recommendations in 1974.[15] Other members of the committee included Oliver, Morris and Yudkin, and Mitchell later gave a revealing account of the manner in which its deliberations were conducted.[16] He described the agreed brief as being to establish whether there was a link between nutrition and cardiovascular disease and, if so, to determine whether modifying the identified risk factor conferred benefit on the individual or community. After many meetings and draft reports, the committee was able to agree on certain conclusions. These were that ischaemic heart disease was multifactorial in origin and that some of the factors involved were dietary in nature but that no single one was predominantly responsible. It also agreed that changes in diet could reduce serum lipids but that there was no certainty that their reduction would decrease susceptibility to heart disease.

The majority of members of the panel accepted the evidence that the death rate from ischaemic heart disease correlated with the saturated fat content of the diet and therefore recommended that the amount of saturated fat in the diet should be reduced. A minority of the members disagreed with this conclusion and felt that this recommendation went beyond the evidence available. However, apart from Yudkin, whose dissent has already been noted, they agreed to support the corporate viewpoint, albeit reluctantly as far as Mitchell was concerned; he undoubtedly belonged to the minority camp on this issue.

In 1984, a second COMA report was published on diet in relation to cardiovascular disease.[17] Tony Mitchell and Michael Oliver were again members of the committee, but it now had several newcomers including the nutritionist Jim Mann, the epidemiologist Geoffrey Rose and the lipidologist Nick Myant. On this occasion the recommendations regarding the need to reduce the saturated fat content of the diet and to increase its ratio of polyunsaturated to saturated fat were far more specific than those made 10 years previously. Nine of the 10 members of the panel concluded that the incidence of coronary heart disease would be reduced by such measures, although they acknowledged that the evidence fell short of proof. The tenth member, who was not named but was undoubtedly Mitchell,[18] believed that the evidence was insufficient to support that conclusion but considered that reducing dietary fat intake might help prevent obesity.

Mitchell amplified his doubts in an editorial review published the same year as the second COMA report.[19] He prefaced it by quoting HL Mencken: "There's always an easy solution to every human problem — neat, plausible and wrong," and went on to question which of the beliefs about diet and coronary disease then current would not only be discarded but also derided by the year 2000. He listed them as follows:

- Coronary heart disease is caused by atherosclerosis.
- Atherosclerotic plaques are cholesterol deposits in artery walls.
- Atherosclerosis can be produced in animals by lipid feeding.
- A high serum cholesterol is a risk marker for coronary disease.
- Dietary lipids determine serum cholesterol.

He went on to point out that if these statements were true, two further ones followed:

- Dietary lipids cause coronary heart disease.
- Dietary modifications will prevent coronary heart disease.

Mitchell then put his head on the block by stating that as far as he was concerned most of these beliefs were untrue or irrelevant (Fig. 9.2). With the benefit of hindsight it is apparent that he was wrong on virtually every count, although this was not quite so obvious at the time, especially regarding the last statement, in that the evidence from diet trials was at best equivocal.

Fig. 9.2.   Tony Mitchell (1928–1991) (Photo: By courtesy of Professor John Hampton).

9.10. Should Every Cow Carry a Government Health Warning?

Mitchell's approach to prevention in the clinic was to tell his patients to stop smoking but to say nothing on the issue of diet. He attributed to Oliver a similarly cautious approach, which led them, he said, to be labelled "the abominable no-men" by "cholesterol evangelists." His own attitude to the latter was equally pejorative. For example, the title of one of his lectures was "Should every cow carry a government health warning?" while his publications had titles such as "Diet and arterial disease — the myths and the realities"[20] and "What constitutes evidence on the dietary prevention of coronary heart disease? Cosy beliefs or harsh facts?"[19] He ended the latter editorial as he had started it, by quoting HL Mencken: "The most costly of all follies is to believe passionately in the palpably-not-true." Or was it even more costly to passionately deny, as he did, what later turned out to be true?

During the late 1970s and early 1980s, McMichael and Mitchell voiced their opposition to the lipid hypothesis in unequivocal terms whereas Oliver adopted a more ambivalent attitude. All three were influential cardiologists and their views undoubtedly affected clinical attitudes among their peers. Mitchell maintained his sceptical attitude to the evidence relating to the prevention of coronary disease right up until his retirement in 1990. By then, the results of the early lipid-lowering drug trials were available, including the Lipid Research Clinics Coronary Primary Prevention trial, which he dismissed as an extravagant waste of money.

9.11. Mitchell: Biography

Tony Mitchell was born in Lancashire in 1928 and studied Medicine at Manchester. After doing his National Service in the Royal Army Medical Corps he was appointed registrar to Sir George Pickering in Oxford. While there he gained his D. Phil by undertaking research into the role of platelets and thrombosis in myocardial infarction and became a keen supporter of the thrombogenic theory of atherosclerosis. From the mid-1960s onwards, in collaboration with John Hampton, he focussed his attention on platelet structure and function, initially in Oxford and then Nottingham. This led them to study factors influencing platelet behaviour and to undertake clinical trials with anti-platelet drugs in patients with myocardial infarction.

Despite his "Dr No" image among supporters of the lipid hypothesis, Tony Mitchell had many positive qualities and was regarded as the major architect of Nottingham's success as a medical school and centre of scientific research. He was an excellent speaker, discussant and chairman and filled the latter role with distinction for several years in the Atherosclerosis Discussion Group (ADG), before it became the British Atherosclerosis Society. He retired from Nottingham in 1990 and died a few months later aged 63 from a pulmonary embolus, which was ironic in that clotting was one of his main scientific interests.[21]

Mitchell was consistent in his demand for unequivocal proof of the lipid hypothesis and, like McMichael, he died before it materialised. The vigour with which they both contested its validity helped ensure that proof of the hypothesis, when it eventually came, was based on the hard evidence they had long demanded.

References

1. Booth C. Sir John McMichael [Obituary]. *Lancet* 1993; **341**: 686.
2. Farewell at Hammersmith. *Br. Med. J.* 1966; **2**: 320.
3. McMichael J. *A Transition in Cardiology: The Mackenzie–Lewis Era. The Harveian Oration of 1975*. London: Royal College of Physicians; 1976.
4. McMichael J. Diet and exercise in coronary heart-disease [Letter]. *Lancet* 1974; **1**: 1340–1341.
5. McMichael J. Prevention of coronary heart-disease [Letter]. *Lancet* 1976; **2**: 569.
6. Shaper G. Prevention of coronary heart-disease [Letter]. *Lancet* 1976; **2**: 1203–1204.
7. McMichael J. Prevention of coronary heart-disease. *Lancet* 1976; **2**: 1350–1351.
8. Goodwin JF. Preventing coronary heart-disease [Letter]. *Lancet* 1977; **1**: 302.
9. McMichael J. Risk factors in coronary heart-disease [Letter]. *Lancet* 1977; **1**: 304.
10. Rose G, Reid DD, McCartney P. Risk factors in coronary heart-disease [Letter]. *Lancet* 1977; **1**: 304.
11. McMichael J. Fats and atheroma: An inquest. *Br. Med. J.* 1979; **1**: 173–175.
12. McMichael J. Fats and arterial disease. *Am. Heart J.* 1979; **98**: 409–412.

13. Mitchell JRA, McMichael J. Letter to the editor. *J. R. Coll. Phys. Lond.* 1979; **13**: 73–74.
14. Mitchell JRA, Schwartz CJ. *Arterial Disease*. Oxford: Blackwell Scientific; 1965.
15. Department of Health and Social Security. Report on Health and Society Subjects. 7. *Diet and Coronary Heart Disease*. Report of the Advisory Panel of the Committee on Medical Aspects of Food Policy (Nutrition) on diet in relation to Cardiovascular and Cerebrovascular Disease. London: HMSO; 1974.
16. Mitchell JRA. Diet and coronary heart disease — A British point of view. *Adv. Exp. Med. Biol.* 1977; **82**: 823–827.
17. Department of Health and Social Security. Report on Health and Society Subjects. 28. *Diet and Cardiovascular Disease*. Committee on Medical Aspects of Food Policy. Report of the Panel on Diet in Relation to Cardiovascular Disease. London: HMSO; 1984.
18. Sanders TAB. Cholesterol, atherosclerosis and coronary disease in the UK, 1950–2000. In: *Wellcome Witnesses to 20th Century Medicine,* Vol. 27, Reynolds LA, Tansey EM (eds.). London: Wellcome Trust Centre for the History of Medicine; 2006.
19. Mitchell JRA. What constitutes evidence on the dietary prevention of coronary heart disease? Cosy beliefs or harsh facts? *Int. J. Cardiol.* 1984; **5**: 287–298.
20. Mitchell JRA. Diet and arterial disease — The myths and the realities. *Proc. Nutr. Soc.* 1985; **44**: 363–370.
21. Mitchell JR. What do we gain by modifying risk factors for coronary disease? *Schweiz. Med. Wochenschr.* 1990; **120**: 359–364.

Chapter 10

Nicolas Myant: Familial Hypercholesterolaemia

10.1. Introduction

"Nature is always hinting at us. It hints over and over again. And suddenly we take the hint." So said Robert Frost although he probably didn't have familial hypercholesterolaemia (FH) in mind. But FH, one of Nature's genetic hints to mankind, is a prime example of the causal role of raised levels of serum cholesterol, specifically of LDL cholesterol, in the pathogenesis of atherosclerosis The fact that the presence of hypercholesterolaemia long precedes the onset of cardiovascular disease (CVD) in affected individuals refutes Wood and McMichael's speculations that a raised serum cholesterol might be the consequence of CVD.

10.2. Background

One of the first scientists to take Nature's hint was Nicolas (Nick) Myant, who had been a house physician to both Sir Thomas Lewis and John McMichael before and after the Second World War, each of them the foremost UK cardiologists of their time. Subsequently, he embarked on a lifelong career with the Medical Research Council (MRC), first at University College Hospital and then at the Hammersmith Hospital, where in 1953 he joined George Popják's MRC Experimental Radiopathology Unit. Myant recalled that Popják was convinced of the validity of the lipid hypothesis and considered that understanding cholesterol synthesis would contribute to the treatment of coronary heart disease[1]: how right he was.

Popják left the Hammersmith in 1962 and Myant became a member of the external staff of the MRC. At that stage, his main research interest was the thyroid gland, including the effect of thyroxine on cholesterol synthesis, and the treatment of thyrotoxicosis with radioactive iodine (^{131}I). In 1963, he was joined by Barry Lewis, who had previously undertaken research on lipids with John Brock in Cape Town and then with Tom Pilkington at St. George's Hospital in London.

After Lewis's arrival at the Hammersmith, Myant's research activities became more lipid-oriented and clinical, the latter facilitated by his being given access to four beds within the hospital. Together they set up an outpatient lipid clinic, the first in Britain, and began to undertake research into the cause and treatment of patients with familial hypercholesterolaemia (FH). Lewis later moved to the Department of Chemical Pathology at the Hammersmith but continued with his research on lipids. Subsequently, he was appointed to the Chair of Chemical Pathology at St. Thomas's Hospital, where he built up a research group which included Gerald Watts and Norman Miller, who, with his brother George, had earlier highlighted the role of HDL cholesterol deficiency as a cardiovascular risk factor.[2]

Myant remained a member of the external staff of the MRC at the Hammersmith until 1969, when space became available in the Cyclotron Unit building. His group was then reconstituted as the MRC Lipid Metabolism Unit (LMU), which the author joined in 1975, and which Myant directed until he retired in 1983. During the earlier part of that period, he was especially interested in the mechanism of hypercholesterolaemia in FH, and he undertook studies of the turnover of ^{14}C-labelled cholesterol to determine whether an abnormality of cholesterol metabolism was responsible.[3]

10.3. Studies of Cholesterol Turnover

Nick's interest in FH dated from 1963 when a young homozygote was referred to him by a physician in Harley Street. Her father was the Iraqi Ambassador to Britain who subsequently became Prime Minister of Iraq until he was imprisoned by Saddam Hussein. At the time of her referral his seven-year-old daughter had a cholesterol level of 24 mmol/l and extensive cutaneous xanthomas. Her hypercholesterolaemia was resistant to diet and drugs so Nick and Barry undertook manual plasmapheresis on four consecutive occasions[1] which lowered her cholesterol but only temporarily and led Nick to comment "This line of treatment was obviously useless." Unfortunately, his comment was equally applicable to the ileal bypass which she underwent subsequently[4] and she died from myocardial ischaemia a few months later, aged 10.

In 1967, Lewis and Myant published the results of their studies of the turnover of ^{14}C-cholesterol in six subjects with probable heterozygous FH and in the patient with homozygous FH described above.[5] Overall, the absolute rate of turnover of cholesterol in the FH patients did not differ from that of the eight control subjects, but it was highest in the homozygote. Myant subsequently stated that the available evidence suggested that over synthesis rather than defective removal of cholesterol was the underlying metabolic defect in FH.[3] With hindsight it is evident that the increase in cholesterol synthesis he observed in that FH homozygote was the consequence and not the cause of the metabolic defect.

10.4. Treatment of Homozygous FH by Apheresis

In 1964, as noted above, Myant and Lewis used manual plasmapheresis to treat a young girl with homozygous FH in a desperate attempt to lower her cholesterol.[1] This involved removing 500 ml of blood on 4 consecutive days, centrifuging it and then returning the red cells to the patient. This resulted in a 37 % decrease in plasma cholesterol from 21 mmol/l, but the level returned to baseline within a few weeks. Soon afterwards, De Gennes *et al.* in Paris adopted a similar approach (they termed it "traitment héroique") in another FH homozygote, who underwent manual plasmapheresis on more than 40 occasions over a period of 3 months; this temporarily reduced his cholesterol but was too onerous to continue permanently.[6] Both these patients had severe atherosclerosis, and their subsequent deaths from cardiovascular disease at the ages of 10 and 23, respectively, were, unfortunately, par for the course in those days.

10.4.1. *Introduction of plasma exchange*

In 1975, the author, Ray Lowenthal — an Australian haematologist at the Hammersmith — and Nick Myant introduced the technique of plasma exchange to lower cholesterol levels in homozygous FH.[7] This novel approach involved using an Aminco continuous flow centrifugal cell separator to exchange cholesterol-free albumin for cholesterol-rich plasma with speed and safety in two FH homozygotes at the Hammersmith Hospital. Performed repeatedly over the years this eventually led to a significant improvement in the longevity of such patients.[8]

The publication in *The Lancet* in May 1975 of the use of plasma exchange to treat FH homozygotes soon had an impact on lipidologists, especially in the USA. During the following months the author was invited to give seminars at the NIH in Bethesda; Baylor College of Medicine in Houston; University of California, San Francisco; University of California, San Diego; and the Massachusetts Institute of Technology in Cambridge (Mass.) and to present a paper in November 1975 at the Annual Meeting of the American Heart Association in Anaheim.

From 1981 onwards, plasma exchange was gradually replaced by lipoprotein apheresis, which initially involved selective removal of LDL by online perfusion of plasma through an immunoadsorbent column.[9] This procedure was in its turn superseded by procedures involving perfusion of plasma or, more recently whole blood, through affinity columns containing either dextran sulphate covalently linked to cellulose beads or poly-acrylate-coated polyacrylamide beads, which adsorb the apolipoprotein B (apoB) component of LDL and of lipoprotein (a) (Lp(a)) and thus removes from the circulation these lipoproteins and their cargo of choles-terol. In the UK, currently the most popular methods involve perfusion of whole blood through dextran sulphate or polyacrylate/polyacrylamide affinity columns.[10]

The main objective of plasma exchange and lipoprotein apheresis was to arrest progression or induce regression of atherosclerosis in FH homozygotes by reducing their very high cholesterol levels and thereby to prolong their lives. Extreme elevations of LDL cholesterol from birth leads to accelerated atherosclerosis, often resulting in valvular and supra-valvular atheroma of the aortic root.[11] Untreated, the average age of death was 18 years,[8] but it could occur in childhood.

10.4.2. *Evidence of clinical benefit from plasma exchange*

The effectiveness of repeated plasma exchange with 2–4 l of albumin as long-term therapy for FH was evaluated in six severely affected patients treated at the Hammersmith.[12] Plasma exchange at monthly intervals for 1 to 2 years reduced mean serum cholesterol levels from 18.5 mmol/l

(715 mg/dl) to 12.4 mmol/l (480 mg/dl) in two female homozygotes but failed to influence xanthomata or prevent a two to threefold increase in their left ventricular aortic systolic pressure gradients.

More effective reduction of cholesterol levels from 15.7 mmol/l (608 mg/dl) to 8.6 mmol/l (333 mg/dl) in two male homozygotes by plasma exchange at fortnightly intervals for 2 to 3 years was accompanied by resolution of xanthomata and by stabilisation of aortocoronary lesions. In two male heterozygotes with angina, coronary angiographic appearances were unaltered or improved after 1 to 2 years of thrice-monthly plasma exchange, which reduced mean serum cholesterol levels from 6.4 mmol/l (248 mg/dl) to 4.7 mmol/l (182 mg/dl). These findings suggested that plasma exchange every 1 to 2 weeks retarded the rate of progression of atheroma in homozygotes and induced regression in heterozygotes.

In 2015, the report of a survey of 44 homozygotes in the UK, of which Myant was a posthumous co-author,[13] analysed clinical outcomes during 50 years of follow up and found that those still alive in 2014 had a lower serum cholesterol than those who had died before that date (8.1 vs. 14.5 mmol/l). They had also sustained fewer major adverse cardiovascular events such as aortic stenosis (33% vs. 77%), reflecting advances in apheresis and drug therapy during the final 20 years covered by the survey. One can conclude that lowering serum cholesterol by plasma exchange initially, and subsequently by lipoprotein apheresis, materially contributed to the improved outlook for FH homozygotes from 1975 onwards. This conclusion was vindicated more recently by the results of a retrospective survey of 133 homozygotes treated in the UK and South Africa with a variety of lipid-lowering therapies including lipoprotein apheresis, which showed that survival was largely determined by the serum cholesterol level while on treatment.[14]

10.5. The South African Contribution to FH

By now it should be obvious that the extraordinarily high levels of LDL cholesterol from birth and premature atherosclerosis that characterise homozygous FH have played an important role in establishing the validity of the lipid hypothesis. The frequency of this rare disorder is greatest in Quebec and South Africa, reflecting inter-marriage within the communities of early French and Dutch settlers, respectively.

In 1980, Harry Seftel described 34 homozygotes he had looked after at the University of Witwatersrand in Johannesburg between 1972–1979, all of them white Afrikaners.[15] He estimated that the frequency of homozygotes in Afrikaners living in the Transvaal was 1:100, a clear example of a founder effect. Having described their phenotypic features he and his colleagues went on to classify South African homozygotes into milder receptor-defective and more severe receptor-negative categories by analysing the LDL-binding properties of their cultured fibroblasts.[16] This was the prelude to a series of DNA-based studies that identified the various LDL receptor mutations responsible for FH in South Africa.

Seftel's protégé and successor Derick Raal subsequently looked after 111 homozygotes in Johannesburg while his compatriots David Marais and Dirk Blom had 38 homozygotes under their care in Cape Town. A genetic diagnosis was made in 86% of the 149 cases, the commonest abnormality being the receptor-defective Afrikaner-1 mutation. In 2011, Raal and Marais analysed the data from their combined cohort to demonstrate that statins lowered LDL cholesterol from its pre-statin baseline by 26% and reduced mortality by 34%.[17] More recently, the Johannesburg and Cape Town workers have pioneered clinical trials of PCSK9 inhibitors in FH heterozygotes and homozygotes and the MTP inhibitor lomitapide in homozygotes, as detailed in Chapter 18. Hence, South African researchers have played an important role during the past 50 years in helping to establish the lipid hypothesis.

10.6. Use of Apheresis as an Investigative Tool

In addition to its therapeutic applications, Myant was struck by the potential of plasma exchange as a means of studying the metabolic effects of acutely perturbing LDL levels. For example, by pre-labelling homozygotes' cholesterol with ^{14}C and comparing the specific activity of the ^{14}C cholesterol in adipose tissue with that in plasma, which initially was 10-fold higher in the former, evidence of an influx of tissue cholesterol into plasma was indicated by a transient rise in the specific activity of plasma cholesterol after each plasma exchange.[7] Subsequent studies in hypercholesterolaemic rabbits subjected to apheresis suggested that the influx of tissue cholesterol was mediated by high-density lipoprotein (HDL).[18]

In another study the non-steady-state turnover of ^{131}I-labelled LDL was determined in four FH patients, three homozygotes and one heterozygote.[19] The fractional and absolute catabolic rates (FCR and ACR) of LDL-apo-B were determined by relating the excretion of radio-activity, measured in urine in vitro and by whole-body counter in vivo, to plasma radioactivity and to LDL specific activity, respectively. The FCR remained relatively constant, even after marked reduction of the expanded LDL pool size by means of plasma exchange. This observation confirmed the existence of an intrinsic defect of LDL catabolism in FH as opposed to saturation of a normal clearance mechanism secondary to overproduction of LDL.

In a third study, the effects on apo-B turnover of markedly reducing plasma LDL concentration by plasma exchange, were studied in four FH patients.[20] Plasma exchange did not increase the rates of synthesis of LDL-apo B in heterozygous or homozygous patients, which supports the hypothesis that LDL production is subject to zero order kinetics i.e. it remains constant irrespective of pool size.

Finally, it was shown that the reduction in plasma LDL cholesterol level achieved by lipoprotein apheresis using a dextran sulphate column in a patient with familial defective apoB (FDB), was similar to that of patients with FH. This confirmed that although LDL apo B in which arginine at position 3500 is replaced by glutamine does not bind to the LDL receptor, it gets removed from plasma by dextran sulphate columns as efficiently as normal LDL.[21]

10.7. ApoB Metabolism in FH

The role of low density lipoprotein (LDL) receptors in the pathogenesis of hereditary and acquired forms of hypercholesterolemia was investigated in vivo by the author and colleagues using a technique devised by Shepherd *et al.* (see Chapter 11). This involved simultaneously determining total and receptor-independent LDL catabolism using ^{125}I-labelled LDL and ^{131}I-labelled LDL coupled with cyclohexanedione.[22] The latter compound inhibits receptor-mediated uptake of LDL.

Receptor-mediated catabolism of LDL, determined as the difference between the turnover of ^{125}I and ^{131}I, was found to be virtually absent in two homozygotes with FH and markedly reduced in a hypothyroid patient. Treatment of the latter with L-thyroxine markedly stimulated receptor-mediated catabolism and reduced her LDL levels. These results demonstrated the existence of an intrinsic and almost total defect of receptor-mediated LDL catabolism in homozygous FH in vivo and revealed an analogous but reversible abnormality in hypothyroidism.

The role of apoB synthesis in the pathogenesis of FH had been investigated in an earlier study by simultaneously measuring the turnovers of ^{125}I-VLDL and ^{131}I-LDL in three homozygotes.[23] VLDL-apoB synthesis rates were normal but LDL-apoB synthesis rates were 1.5–2-fold greater than VLDL-apoB synthesis rates, suggesting direct secretion of LDL into plasma occurs in homozygous FH, presumably by the liver.

10.8. Familial Defective ApoB (FDB)

After he retired as Director of the MRC LMU, Nick Myant did not retire from scientific research — far from it. His biochemical colleague Anne Soutar found some bench space for him in her laboratory where he and his young research assistant, John Gallagher, undertook a series of fruitful investigations into FDB. Anne herself had already started on an equally fruitful series of investigations of the genetic mutations of FH[24] and she provided Nick and John with the tools of molecular biology.

In 1990, Nick and his co-authors identified 10 individuals with FDB by screening hyperlipidaemic subjects in the UK and Scandinavia.[25] They found that the frequency of the mutation in persons with an FH phenotype was 3%. Clinical features and plasma lipid levels of FDB subjects were similar to those of FH heterozygotes. Subsequently Gallagher and Myant identified two patients with homozygous FDB and showed that the binding affinity of their LDL for the LDL (apoB/E) receptor was only 10–20% of normal whereas the binding efficiency of their VLDL remnants was normal.[26] This was attributed to apoE-mediated binding of VLDL remnants to LDL receptors, which function normally in FDB,[27] and presumably accounts for the fact that the phenotypic expression of this disorder is milder than that of classical receptor-deficient FH. Because LDL receptor-mediated removal of VLDL remnants is normal in FDB, this reduces their conversion to LDL in plasma, thereby decreasing LDL production.

Myant's final contribution to the literature on FDB was in 1997, when he celebrated his 80th birthday by publishing his estimate of the age of the mutation in apoB that was responsible for the disorder.[28] He calculated that the apoB R3500Q mutation responsible for FDB occurred in the human apoB gene in Europe approximately 6,000–7,000 years or

270 generations ago, based on painstaking genotyping of samples obtained from FDB families in various parts of the world.

10.9. Biography

Nick Myant was born in Cardiff in 1917, the son of a Belgian father and English mother. After attending public school he studied Medicine at Balliol College, Oxford and University College Hospital (UCH), London, where he was house physician to Sir Thomas Lewis. Between 1944 and 1946, he served in the army in India and afterwards became house physician to Professor John McMichael at the Hammersmith Hospital. In 1953 he joined the MRC Experimental Radiopathology Research Unit at the Hammersmith, directed by George Popják. One of Popják's major interests was the enzymology of cholesterol metabolism and it was this that kick-started Nick's interest in cholesterol.

As discussed earlier, analysis of the results of whole-body cholesterol turnover studies in normal and FH subjects led Nick to conclude in 1970 that over-synthesis, rather than defective removal of cholesterol, was the underlying abnormality in FH. This issue was addressed at an international conference organised by Nick and the author at the Hammersmith in 1975, attended by virtually all the world leaders in the field. Popják, who by then had moved to UCLA, supported the view that overproduction of cholesterol (resulting from dysregulation of HMG-CoA reductase) was the primary defect in FH, rather than defective removal, which prompted a heated discussion between various members of the audience, including Brown and Goldstein.

Subsequently, after examining the evidence for the competing hypotheses, Nick became convinced that the "excess production" theory was wrong and he accepted the "defective removal" theory, which was based on the discovery of the LDL receptor the previous year by Brown and Goldstein. It followed, therefore, that the excessive production of cholesterol in FH was secondary to defective cellular uptake of cholesterol and the resulting lack of feedback control. Nick greatly admired the Dallas duo and their Nobel Prize-winning research (Fig. 10.1). This admiration was reciprocated and after Nick's death Michael Brown wrote in an email to the author: "When we entered the field we found him to be one of a very small number of linear thinkers. The rest were running around in circles. We held him in the highest regard."

Fig. 10.1.   Nick Myant (1917–2015).

The MRC established the LMU at Hammersmith Hospital in 1969 with Nick as director and during the ensuing years Nick gained the respect and affection of all who worked with him. By the time he retired in 1983, his Unit had made significant contributions towards a better understanding of the metabolic basis, genetic diagnosis and management of FH, as described in this chapter. But Nick didn't really retire. From 1984–1987 he co-edited *Atherosclerosis* and then embarked on a second career, this time in molecular biology, using techniques which he learned at the Dunn School of Pathology at Oxford and then brought back to Hammersmith.

Nick's pre-eminent role in lipid research in Britain is commemorated by the Myant Lecture, which was inaugurated at the scientific meeting of the British Hyperlipidaemia Association (now HEART UK) in 1989 and has been given annually thereafter apart from 2020, when COVID prevented it. Many renowned lipidologists have been Myant Lecturers including Dan Steinberg, Joe Goldstein, Dick Havel and others (Table 10.1). This eponymous lecture is a fitting tribute to a British clinical scientist who ranks alongside those American pioneers of lipid

Table 10.1.   Myant Lecturers, 1989–2022.

British Hyperlipidaemia Association	HEART UK
1989 N. Miller	2003 J. Chapman
1990 D. Steinberg	2004 P. Barter
1991 A. Sniderman	2005 S. Haffner
1992 J. Goldstein	2006 R. Hegele
1993 S. Grundy	2007 J. Kastelein
1994 R. Havel	2008 C. Sirtori
1995 A. Gotto	2009 J. Shepherd
1996 R. Lawn	2010 P. Durrington
1997 G. Assmann	2011 E. Schaefer
1998 G. Thompson	2012 K. Frayn
1999 J. Breslow	2013 D. Marais
2000 B. Brewer	2014 A. von Eckardstein
2001 L. Chan	2015 S. Humphries
2002 F. Sacks	2016 T. Sanders
	2018 B. Staels
	2019 S. O'Rahilly
	2020 No conference
	2021 A. Neil
	2022 F. Raal

research. It also perpetuates the memory of a unique colleague — a friendly, charming and unassuming man with strong left-wing views, whose formidable intelligence was enlivened by a quirky sense of humour.

It has been said that there are two types of ambitious people. There are people who want to *do* something and there are people who want to *be* somebody. Nick was very much a scientist of the first type.[29] He died in 2015 aged 97.

References

1. Myant NB. Plasma cholesterol as a cause of coronary heart disease (CHD); the cholesterol-CHD hypothesis. In: *Wellcome Witnesses to 20th Century Medicine,* Vol. 27, Reynolds LA, Tansey EM (eds.). London: Wellcome Trust Centre for the History of Medicine; 2006.

2. Miller GJ, Miller NE. Plasma-high-density-lipoprotein concentration and development of ischaemic heart-disease. *Lancet* 1975; **1**: 16–20.
3. Myant NB. The regulation of cholesterol metabolism as related to familial hypercholesterolaemia. *Sci. Basis Med. Annu. Rev.* 1970; 230–259.
4. Johnston ID, Davis JA, Moutafis CD, Myant NB. Ileal bypass in the management of familial hypercholesterolaemia. *Proc. R Soc. Med.* 1967; **60**: 16–18.
5. Lewis B, Myant NB. Studies in the metabolism of cholesterol in subjects with normal plasma cholesterol levels and in patients with essential hypercholesterolaemia. *Clin. Sci.* 1967; **32**: 201–213.
6. De Gennes JL, Touraine R, Maunand B, *et al.* Homozygous cutaneotendinous forms of hypercholesteremic xanthomatosis in an exemplary familial case. Trial of plasmapheresis: An heroic treatment. *Bull. Mem. Soc. Med. Hop. Paris* 1967; **118**: 1377–1402.
7. Thompson GR, Lowenthal R, Myant NB. Plasma exchange in the management of homozygous familial hypercholesterolaemia. *Lancet* 1975; **1**(7918): 1208–211.
8. Thompson GR, Miller JP, Breslow JL. Improved survival of patients with homozygous familial hypercholesterolaemia treated with plasma exchange. *Br. Med. J.* 1985; **291**: 1671–1673.
9. Stoffel W, Borberg H, Greve V. Application of specific extracorporeal removal of low density lipoprotein in familial hypercholesterolaemia. *Lancet* 1981; **2**(8254): 1005–1007.
10. Pottle A, Thompson G, Barbir M, *et al.* Lipoprotein apheresis efficacy, challenges and outcomes: A descriptive analysis from the UK Lipoprotein Apheresis Registry, 1989–2017. *Atherosclerosis* 2019; **290**: 44–51.
11. Allen JM, Thompson GR, Myant NB, Steiner R, Oakley CM. Cardiovascular complications of homozygous familial hypercholesterolaemia. *Br. Heart J.* 1980; **44**: 361–368.
12. Thompson GR, Myant NB, Kilpatrick D, *et al.* Assessment of long-term plasma exchange for familial hypercholesterolaemia. *Br. Heart J.* 1980; **43**: 680–688.
13. Thompson GR, Seed M, Naoumova RP, *et al.* Improved cardiovascular outcomes following temporal advances in lipid-lowering therapy in a genetically characterised cohort of familial hypercholesterolaemia homozygotes. *Atherosclerosis* 2015; **243**: 328–333.
14. Thompson GR, Blom DJ, Marais AD, *et al.* Survival in homozygous familial hypercholesterolaemia is determined by the on-treatment level of serum cholesterol. *Eur. Heart J.* 2018; **39**: 1162–1168.
15. Seftel HC, Baker SG, Sandler MP, *et al.* A host of hypercholesterolaemic homozygotes in South Africa. *Br. Med. J.* 1980; **281**: 633–636.
16. van der Westhuyzen DR, Coetzee GA, Demasius IP, *et al.* Low density lipoprotein receptor mutations in South African homozygous familial hypercholesterolemic patients. *Arteriosclerosis* 1984; **4**: 238–247.

17. Raal FJ, Pilcher GJ, Panz VR, *et al.* Reduction in mortality in subjects with homozygous familial hypercholesterolemia associated with advances in lipid-lowering therapy. *Circulation* 2011; **124**: 2202–2207.
18. Kano M, Koizumi J, Jadhav A, Thompson GR. Plasma exchange and low density lipoprotein apheresis in Watanabe heritable hyperlipidemic rabbits. *Arteriosclerosis* 1987; **7**: 256–261.
19. Thompson GR, Spinks T, Ranicar A, Myant NB. Non-steady-state studies of low-density-lipoprotein turnover in familial hypercholesterolaemia. *Clin. Sci. Mol. Med.* 1977; **52**: 361–369.
20. Soutar AK, Myant NB, Thompson GR. Metabolism of apolipoprotein B-containing lipoproteins in familial hypercholesterolaemia: Effects of plasma exchange. *Atherosclerosis* 1979; **32**: 315–325.
21. Maher VM, Kitano Y, Neuwirth C, *et al.* Effective reduction of plasma LDL levels by LDL apheresis in familial defective apolipoprotein B-100. *Atherosclerosis* 1992; **95**: 231–234.
22. Thompson GR, Soutar AK, Spengel FA, *et al.* Defects of receptor-mediated low density lipoprotein catabolism in homozygous familial hypercholesterolemia and hypothyroidism in vivo. *Proc. Natl. Acad. Sci. USA* 1981; **78**: 2591–2595.
23. Soutar AK., Myant NB, Thompson GR. Simultaneous measurement of apolipoprotein B turnover in very low and low density lipoproteins in familial hypercholesterolaemia. *Atherosclerosis* 1977; **28***: 247–256.
24. Soutar AK, Knight BL, Patel DD. Identification of a point mutation in growth factor repeat C of the low density lipoprotein-receptor gene in a patient with homozygous familial hypercholesterolemia that affects ligand binding and intracellular movement of receptors. *Proc. Natl. Acad. Sci. USA* 1989; **86**: 4166–4170.
25. Tybjaerg-Hansen A, Gallagher J, Vincent J, *et al.* Familial defective apolipoprotein B-100: Detection in the United Kingdom and Scandinavia, and clinical characteristics of ten cases. *Atherosclerosis* 1990; **80**: 235–242.
26. Gallagher JJ, Myant NB. The affinity of low-density lipoproteins and of very-low-density lipoprotein remnants for the low-density lipoprotein receptor in homozygous familial defective apolipoprotein B-100. *Atherosclerosis* 1995; **115**: 263–272.
27. Myant NB. Familial defective apolipoprotein B-100: A review, including some comparisons with familial hypercholesterolaemia. *Atherosclerosis* 1993; **104**: 1–18.
28. Myant NB, Forbes SA, Day IN, Gallagher J. Estimation of the age of the ancestral arginine$_{3500}$-->glutamine mutation in human apoB-100. *Genomics* 1997; **45**: 78–87.
29. Gibbons G, Soutar A, Thompson G. In Memoriam: Nick Myant (1917–2015). *J. Lipid Res.* 2015; **56**: 1081.

Chapter 11

Joseph Goldstein and Michael Brown: Discovery of the LDL Receptor[*]

11.1. Introduction

In 1985, the Nobel Prize for Physiology or Medicine was awarded to Michael Brown and Joseph Goldstein, then aged 44 and 45, respectively, "for their discoveries concerning the regulation of cholesterol metabolism." In his presentation speech, Professor Viktor Mutt of the Karolinska Institute commented that they had discovered "a physiological mechanism of great importance: the way in which mammalian cells … establish an equilibrium between their own synthesis and the cholesterol they obtain from circulating blood" and that "this knowledge forms a rational basis for development of methods for the treatment and prevention of the widespread disabling diseases known to be a consequence of derangement in plasma cholesterol concentrations."[1]

11.2. Biographical Background of the Laureates

11.2.1. *Joseph Goldstein*

Joseph L. Goldstein was born in 1940 and grew up in South Carolina, USA. He attended the local schools in his hometown and then went to

---

[*]Edited from: Thompson GR. Discovery of the LDL receptor and its role in cholesterol metabolism. In: *Nobel Prizes that changed Medicine*, Thompson G (ed.). London: Imperial College Press; 2012.

Washington and Lee University in Virginia, where he graduated *summa cum laude* in Chemistry in 1962. Having decided to study Medicine he gained entry to the University of Texas Southwestern Medical School in Dallas. His talents were soon spotted by Donald Seldin, legendary head of the Department of Medicine, who ear-marked him as the future head of a division of Medical Genetics. When Goldstein obtained his MD in 1966, he was acknowledged as the outstanding graduate of his year. He spent the next 2 years as an intern and resident at the Massachusetts General Hospital (MGH) in Boston, where the author was working as a research fellow at the time.

Joe came over as a serious and intense yet open and friendly person, who spoke in rapid sentences with a pronounced Southern accent. After completing his residency, he spent the next 2 years as a clinical associate at the National Institutes of Health (NIH) in Bethesda, where he looked after Don Fredrickson's patients, including some with homozygous familial hypercholesterolaemia (FH).

Following Seldin's advice he went to Seattle in 1970 to do a fellowship in Medical Genetics with Arno Motulsky at the University of Washington. During his 2 years there he undertook a major study of the genetic causes of hyperlipidaemia in myocardial infarction survivors.[2–4] He returned to Dallas in 1972 and was appointed Assistant Professor in charge of the division of Medical Genetics and promoted Professor in 1976. The following year he became Chairman of the Department of Molecular Genetics and has remained so ever since.

11.2.2. *Michael Brown*

Michael S. Brown was born in 1941 in Brooklyn, New York, and moved to Pennsylvania when he was eleven. He went to high school in a small town near Philadelphia and thence to the University of Pennsylvania, where he graduated with a degree in Chemistry in 1962. He went on to the University of Pennsylvania School of Medicine and got his MD there in 1966, coming top of the class like Joe Goldstein had done in Dallas. The two met later that year when they were interns at the MGH and quickly became friends.

In 1968, Brown went to the NIH as a Clinical Associate in gastroenterology and spent some time in Earl Stadtman's laboratory, where he acquired skills in enzymology. In 1971, he embarked upon a fellowship in gastroenterology at the University of Texas Southwestern Medical School, having been persuaded to move there by Joe Goldstein. In addition to undertaking clinical duties, Mike Brown worked in John Dietschy's laboratory where the two of them, together with Marvin Siperstein, succeeded in solubilising and partially purifying from rat liver the enzyme hydroxyl methyl glutaryl co-enzyme A (HMG CoA) reductase, the rate limiting step on the cholesterol synthesis pathway.[5] He became Associate Professor in Internal Medicine in 1974, Professor in 1976 and Professor of Medical Genetics and Director of the Center for Genetic Diseases the following year.

Scientific collaboration between Brown and Goldstein began in 1972, after the latter had returned to Dallas from Seattle, and was precipitated by a phone call from the surgeon Tom Starzl, informing Siperstein that he was about to perform a portocaval shunt on a 12-year-old girl with homozygous FH in an attempt to lower her cholesterol. Siperstein was away so Mike Brown flew to Denver to collect a skin biopsy from the patient[6]; the research that ensued is described in what follows and formed the scientific basis for their Nobel Prize. At a personal level Brown appeared less restless than Goldstein and had a more measured approach to life. Both are superb communicators and have been the joint recipients of numerous honours and prizes, including the Lasker Award for Basic Medical Research, which they received in the same year as their Nobel Prize.

Goldstein and Brown were doing their internships in Boston in 1967 when Fredrickson, Levy and Lees published their five-part article in the *New England Journal of Medicine* introducing a novel classification of disorders of lipoprotein metabolism.[7] The article aroused enormous interest among clinicians interested in the role of lipids in atherosclerosis and cardiovascular disease. It also underlined the unique status of the NIH, where Fredrickson directed the Laboratory of Molecular Diseases in the National Heart Institute, as a referral centre for inherited forms of dyslipidaemia including patients with familial hypercholesterolaemia, a disorder which fascinated Joe Goldstein when he worked there.

11.3. Familial Hypercholesterolaemia (FH)

Familial hypercholesterolaemia is characterised by hypercholesterolaemia from birth and the subsequent development of tendon xanthomas and premature onset of atherosclerosis, as first described by Muller in 1939.[8] Myant noted that the dominantly inherited increase in plasma cholesterol was largely confined to LDL cholesterol,[9] which can reach >20 mmol/l in homozygotes and up to 12 mmol/l in heterozygotes. In homozygotes the severe hypercholesterolaemia usually results in atheromatous involvement of the aortic root before puberty. Sudden death from myocardial infarction or acute coronary insufficiency before the age of 30 was the rule in untreated patients. At post-mortem examination the aortic valve, sinuses of Valsalva and ascending arch of the aorta are grossly infiltrated with atheroma, with the coronary ostia often narrowed to pinhole size.

Then as now, heterozygous FH often remained undiagnosed until the onset of cardiovascular symptoms in adult life. In addition to hypercholesterolaemia there were signs of cholesterol deposition, such as corneal arcus and tendon xanthomas. The increased frequency and premature onset of coronary heart disease is well documented and, in untreated subjects, occurred about 20 years earlier than in the rest of the population.

This then was the disorder which so intrigued Goldstein and Brown in 1972. Because hypercholesterolaemia was present in heterozygotes as well as in homozygotes, they speculated that FH was probably not caused by an enzyme deficiency, as is common in recessively inherited disorders, but was more likely to be due to a defect in a protein involved in the feedback regulation of cholesterol synthesis.[10] They tested this hypothesis in a series of ingenuous experiments using fibroblasts cultured from the skin of FH patients and normal subjects.

11.4. Discovery of the LDL Receptor and the Cause of FH

Working together in Dallas, combining Goldstein's cell culture skills and Brown's knowledge of enzymology, the pair set out to measure the activity of HMG CoA reductase in fibroblasts cultured from the skin of the FH homozygote whose serum cholesterol had reached 26 mmol/l and upon whom Starzl had operated.[11] They discovered that the activity of HMG CoA reductase was 60–80 times greater in her fibroblasts when incubated with lipoprotein-deficient plasma than in control fibroblasts. And unlike the latter, HMG CoA reductase activity in the homozygote's fibroblasts, determined by measuring the conversion of ^{14}C-HMG CoA to ^{14}C-mevalonate, was not suppressed by adding LDL to the culture medium. Fibroblasts from FH heterozygotes exhibited a partial defect in regulation of HMG CoA reductase, but Goldstein and Brown considered it unlikely that a defect in the gene encoding this enzyme could be the cause of FH.

Further investigations demonstrated that [125] I-labelled LDL was bound to normal fibroblasts by a high affinity, saturable process which mediated the uptake and subsequent proteolytic degradation of LDL and resulted in suppression of HMG CoA reductase activity and therefore of cholesterol synthesis.[12] They showed that fibroblasts from FH homozygotes lacked these high affinity binding sites, failed to degrade LDL and failed to suppress their HMG CoA reductase activity when exposed to LDL. Fibroblasts from FH heterozygotes showed intermediate levels of high affinity binding of LDL.

Goldstein and Brown concluded that high affinity binding of LDL results in endocytosis (internalisation) and lysosomal degradation of LDL, the ensuing hydrolysis of cholesterol esters leading to release of free cholesterol and inhibition of HMG CoA reductase. They proposed that FH was due to an abnormality of a dominantly inherited gene whose product is the LDL receptor, defects of which result in reduced catabolism of LDL and increased cholesterol synthesis. The severity of the ensuing hypercholesterolaemia depended upon whether the receptor defect was partial, as in heterozygotes, or total, as in homozygotes.

Evidence that LDL catabolism was defective in FH *in vivo* came from several sources. Turnover studies at the NIH had previously shown that the fractional catabolic rate (FCR) of ^{125}I-labelled LDL in FH heterozygotes was only half that of normal subjects.[13] Non-steady state studies of LDL turnover in FH patients showed that the FCR of LDL remained subnormal even after a marked reduction in pool size following plasma exchange, suggesting that the catabolic defect was intrinsic and not secondary to the expanded pool.[14] Subsequent studies by Shepherd *et al.*[15] in Glasgow involved simultaneous administration of native LDL labelled with ^{125}I and ^{131}I-labelled LDL coupled with 1,2 cyclohexanedione, a chemical modification known to inhibit binding of LDL to fibroblasts.[16] Their results showed that the FCR of ^{125}I-labelled LDL but not that of ^{131}I-LDL-cyclohexanedione was reduced in FH heterozygotes compared with normal subjects, implying decreased receptor-mediated catabolism of LDL in FH. Using the same technique the author and his colleagues in London demonstrated an analogous but more severe defect in homozygotes.[17]

The likelihood that the abnormalities of receptor-mediated catabolism revealed by these turnover studies reflected impaired uptake of LDL by hepatic receptors was supported by *in vitro* binding studies, using cell membranes prepared from liver biopsies obtained during routine abdominal surgery from control subjects and FH patients.[18] Saturable binding of ^{125}I-LDL was reduced by approximately 50% in FH heterozygotes compared with normal subjects; when the data were pooled they showed an inverse correlation between the saturable binding of LDL by liver membranes and the concentration of cholesterol in plasma. This finding drew attention to the importance of LDL receptors in the liver in regulating plasma cholesterol, a premise which was later confirmed after Starzl performed a liver transplant in an FH homozygote and thereby restored her plasma cholesterol level to normal.[19]

Bilheimer, Grundy and their Dallas colleagues went on to show that the turnover rate of labelled LDL was grossly subnormal in the FH homozygote pre-transplant and was restored to near normal post-transplant, confirming the importance of the liver in LDL catabolism. Scott Grundy subsequently played a major role in developing the Adult Treatment Panel guidelines of the National Cholesterol Education Program. Hence, his collaborative studies with Brown and Goldstein enabled him to combine undertaking clinical research in individual patients with the development later on of highly influential strategies for the prevention of cardiovascular disease at the national level.[20]

11.5. Analysis of the LDL Receptor Gene and Mutations Causing FH

In 1982, Goldstein and Brown and their colleagues succeeded in purifying the LDL receptor from bovine adrenal cortex, yielding a glycoprotein with a molecular weight of 164,000.[21] The purified receptor had LDL binding properties which were identical to those of receptors expressed on intact cells. Two years later they cloned and sequenced the cDNA of the human LDL receptor.[22] The latter consisted of a protein containing 839 amino acids grouped into five domains: an amino terminal domain containing the binding site for LDL; a domain homologous with the precursor of the mouse epidermal growth factor receptor; a domain adjacent to the plasma membrane which is the site of O-linked glycosylation; a membrane spanning domain that anchors the receptor to the cell surface; and a

Fig. 11.1.   Michael Brown and Joseph Goldstein at the 10th anniversary celebration of their Nobel Prize in Stockholm.

cytoplasmic carboxy terminal. They isolated the LDL receptor gene the following year and showed that it contained 18 exons.[23] The discovery and characterisation of the LDL receptor gave a unique insight into the pathophysiology of cholesterol metabolism and was the main scientific advance for which Goldstein and Brown were subsequently awarded their Nobel Prizes and Lasker Awards (Fig. 11.1).

These advances paved the way for analysis of the various different mutations of the LDL receptor gene which cause FH and well over a thousand of these have been described to date. The nature of the underlying mutation can exert a marked influence on the severity of the clinical phenotype. For example, FH heterozygotes with a mutation in exon 4, which encodes a critical part of the receptor's LDL-binding domain, have higher LDL cholesterol levels than those with mutations in other parts of the gene which do not affect this domain.[24] Soutar and colleagues explained these findings by showing that lymphoblastoid cells cultured from FH patients with "severe" mutations that cause receptor deficiency degrade less LDL than do cells from patients with "mild" mutations, which are often single amino acid substitutions that enable LDL receptors to be expressed but impair their ability to bind LDL.[25]

Several years earlier Brown and Goldstein[26] had investigated a homozygote with a most unusual mutation and the results of their studies provided an important insight into the process whereby LDL undergoes endocytosis. Fibroblasts from this patient (J.D.) were able to bind ^{125}I-LDL, unlike the fibroblasts from other homozygotes, but differed from normal fibroblasts in their inability to transport LDL into the cell and inhibit HMG CoA reductase. Subsequent studies revealed that the failure of J.D.'s receptors to internalise LDL was due to their inability to cluster in clathrin-coated pits on the cell surface[27] whereas in normal fibroblasts LDL binds to receptors within these pits, which then invaginate to form endocytic vesicles and deliver LDL to lysosomes. More recent studies revealed that the mutation responsible causes a single amino acid substitution, of cysteine for tyrosine, in the cytoplasmic domain of the LDL receptor[28] and that this prevents receptors from clustering in coated pits. The unravelling of this defect gave important insights into normal LDL metabolism and provides a classic example of the meticulous scientific detective work which has emanated from the Goldstein and Brown laboratory in Dallas for the past 50 years.

11.6. Impact of These Discoveries on Medical Science

The award of the Nobel Prize to Goldstein and Brown in 1985 helped boost the morale of those working in the field of lipids and atherosclerosis at a time when there were serious doubts regarding the relevance of cholesterol to coronary disease. Their discovery of the LDL receptor provided a novel mechanism for the physiological uptake by cells of cholesterol transported in LDL and for the latter's catabolism by the liver. Equally important, it not only established the nature of the genetic defect in familial hypercholesterolaemia, and thus led to DNA analysis becoming an important screening test for the disease, but it re-emphasised the causal nature of the link between LDL cholesterol and atherosclerosis.

Goldstein and Brown's subsequent discovery of the scavenger receptor resulted in increased understanding of the pathogenesis of atherosclerosis and the key role played by modified LDL in this process. Finally, together with Endo, they defined the mechanism of action of the first HMG CoA reductase inhibitor compactin in inhibiting cholesterol synthesis and upregulating LDL receptors, thereby contributing to the development of subsequent statins. The introduction of these compounds into clinical practice in 1987 resulted in dramatic improvements in the prognosis of patients with FH[29,30] and in the management and prevention of cardiovascular disease throughout the world.[31]

References

1. *Nobel Lectures, Physiology or Medicine 1981–1990*, Lindsten J. (ed.). Singapore: World Scientific Publishing Co.; 1993.
2. Goldstein JL, Hazzard WR, Schrott HG, Bierman EL, Motulsky AG. Hyperlipidemia in coronary heart disease. I. Lipid levels in 500 survivors of myocardial infarction. *J. Clin. Invest.* 1973; **52**: 1533–1543.
3. Goldstein JL, Schrott HG, Hazzard WR, Bierman EL, Motulsky AG. Hyperlipidaemia in coronary heart disease. II. Genetic analysis of lipid levels in 176 families and delineation of a new inherited disorder, combined hyperlipidemia. *J. Clin. Invest.* 1973; **52**: 1544–1568.
4. Hazzard WR, Goldstein JL, Schrott MG, Motulsky AG, Bierman EL. Hyperlipidaemia in coronary heart disease. III. Evaluation of lipoprotein phenotypes of 156 genetically defined survivors of myocardial infarction. *J. Clin. Invest.* 1973; **52**: 1569–1577.
5. Brown MS, Dana SE, Dietschy JM, *et al*. 3-hydroxy-3-methylglutaryl coenzyme A reductase. Solubilization and purification of a cold-sensitive microsomal enzyme. *J. Biol. Chem.* 1972; **248**: 4731–4738.
6. Foster DW, Wilson JD. Presentation of the Kober Medal to Joseph L. Goldstein and Michael S. Brown. *J. Clin. Invest.* 2002; **110**: S5–9.
7. Fredrickson DS, Levy RI, Lees RS. Fat transport in lipoproteins — An integrated approach to mechanisms and disorders. *N. Engl. J. Med.* 1967; **276**: 34–42, 94–103, 148–156, 215–225, 273–281.
8. Muller C. Angina pectoris in hereditary xanthomatosis. *Arch. Intern. Med.* 1939; **64**: 675–700.
9. Myant NB. The metabolic lesion in familial hypercholesterolaemia. In: *Cholesterol Metabolism and Lipolytic Enzymes*, Polonovski J. (ed.). New York: Masson Publishing; 1977.
10. Goldstein JL, Brown MS. The LDL receptor. *Arterioscler. Thromb. Vasc. Biol.* 2009; **29**: 431–438.
11. Goldstein JL, Brown MS. Familial hypercholesterolemia: Identification of a defect in the regulation of 3-hydroxy-3-methylglutaryl coenzyme A

reductase activity associated with overproduction of cholesterol. *Proc. Natl. Acad. Sci. USA* 1973; **70**: 2804–2808.

12. Goldstein JL, Brown MS. Binding and degradation of low density lipoproteins by cultured human fibroblasts. *J. Biol. Chem.* 1974; **249**: 5153–5162.

13. Langer T, Strober W, Levy RI. The metabolism of low density lipoprotein in familial type II hyperlipoproteinemia. *J. Clin. Invest.* 1972; **51**: 1528–1536.

14. Thompson GR, Spinks T, Ranicar A, *et al*. Non-steady-state studies of low density lipoprotein turnover in familial hypercholesterolaemia. *Clin. Sci. Mol. Med.* 1977; **52**: 361–369.

15. Shepherd J, Bicker S, Lorimer AR, *et al*. Receptor-mediated low density lipoprotein catabolism in man. *J. Lipid Res.* 1979; **20**: 999–1006.

16. Mahley RW, Innerarity TL, Pitas RE, *et al*. Inhibition of lipoprotein binding to cell surface receptors of fibroblasts following selective modification of arginyl residues in arginine-rich and B apoproteins. *J. Biol. Chem.* 1977; **252**: 7279–7287.

17. Thompson GR, Soutar AK, Spengel FA, *et al*. Defects of receptor-mediated low density lipoprotein catabolism in homozygous familial hypercholesterolemia and hypothyroidism *in vivo*. *Proc. Natl. Acad. Sci. USA* 1981; **78**: 2591–2595.

18. Harders-Spengel K, Wood CB, Thompson GR, *et al*. Difference in saturable binding of low density lipoprotein to liver membranes from normocholesterolemic subjects and patients with heterozygous familial hypercholesterolemia. *Proc. Natl. Acad. Sci. USA* 1982; **79**: 6355–6359.

19. Bilheimer DW, Goldstein JL, Grundy SM, *et al*. Liver transplantation to provide low-density-lipoprotein receptors and lower plasma cholesterol in a child with homozygous familial hypercholesterolemia. *N. Eng. J. Med.* 1984; **311**: 1658–1664.

20. Grundy S, Brown WV. From the editor: An interview with Dr. Scott Grundy. *J. Clin. Lipidol.* 2014; **8**: 1–8.

21. Schneider WJ, Beisiegel U, Goldstein JL, *et al*. Purification of the low density lipoprotein receptor, an acidic glycoprotein of 164,000 molecular weight. *J. Biol. Chem.* 1982; **257**: 2664–2673.

22. Yamamoto T, Davis CG, Brown MS, *et al*. The human LDL receptor: A cysteine-rich protein with multiple Alu sequences in its mRNA. *Cell* 1984; **39**: 27–38.

23. Sudhof TC, Goldstein JL, Brown MS, *et al*. The LDL receptor gene: A mosaic of exons shared with different proteins. *Science* 1985; **228**: 815–822.

24. Gudnason V, Day IN, Humphries SE. Effect on plasma lipid levels of different classes of mutations in the low-density lipoprotein receptor gene in

patients with familial hypercholesterolemia. *Arterioscler. Thromb.* 1994; **14**: 1717–1722.

25. Xi-Ming Sun, Patel DP, Knight BL, *et al.* Comparison of the genetic defect with LDL-receptor activity in cultured cells from patients with a clinical diagnosis of heterozygous familial hypercholesterolemia. *Arterioscler. Thromb. Vasc. Biol.* 1997; **17**: 3092–3101.

26. Brown MS, Goldstein JL. Analysis of a mutant strain of human fibroblasts with a defect in the internalization of receptor-bound low density lipoprotein. *Cell* 1976; **9**: 663–674.

27. Anderson RG, Goldstein JL, Brown MS. A mutation that impairs the ability of lipoprotein receptors to localise in coated pits on the cell surface of human fibroblasts. *Nature* 1977; **270**: 695–699.

28. Davis CG, Lehrman MA, Russell DW, *et al.* The J.D. mutation in familial hypercholesterolemia. Amino acid substitution in cytoplasmic domain impedes internalization of LDL receptors. *Cell* 1986; **45**: 15–24.

29. Neil A, Cooper J, Betteridge J, *et al.* Reductions in all-cause, cancer, and coronary mortality in statin-treated patients with heterozygous familial hypercholesterolaemia: A prospective registry study. *Eur. Heart J.* 2008; **29**: 2625–2633.

30. Versmissen J, Oosterveer DM, Yazdanpanah M, *et al.* Efficacy of statins in familial hypercholesterolaemia: A long- term cohort study. *BMJ* 2008; **337**: a2423.

31. https://laskerfoundation.org/winners/statins-for-lowering-ldl-and-decreasing-heart-attacks/.

Chapter 12

Daniel Steinberg: The Role of Oxidised LDL in Atherosclerosis

12.1. Introduction: The Scavenger Receptor

Goldstein and Brown showed that under normal circumstances the LDL receptor is the means by which various types of cells such as hepatocytes acquire cholesterol for their metabolic needs. The observation that cholesterol ester-filled macrophages or foam cells occurred in the atherosclerotic lesions of FH homozygotes lacking LDL receptors intrigued them, since it implied that macrophages took up LDL cholesterol via an LDL receptor-independent pathway. Using mouse peritoneal macrophages they showed that the latter bound and degraded chemically synthesised acetyl LDL much faster than native LDL and continued to do so despite accumulating large amounts of cholesterol.[1] They demonstrated that this process reflected the presence of a high affinity binding site on macrophages, which recognised acetylated but not native LDL.

Brown *et al.* also showed that uptake and degradation of acetylated LDL markedly stimulated cholesterol esterification in mouse peritoneal macrophages and greatly increased their cholesterol ester content.[2] The latter reflected the activity of acylcholesterol acyltransferase (ACAT) within macrophages, which re-esterified cholesterol linoleate, the major form of esterified cholesterol in LDL, to cholesterol oleate, the form in which it accumulates in foam cells in atherosclerotic lesions. Human monocyte macrophages behaved in a similar manner and Brown *et al.* suggested that some chemical or physical modification of LDL occurred in plasma or tissue fluid which made it a suitable ligand for what they termed a scavenger pathway for LDL in FH. Subsequently, it became clear that this pathway plays a key role in atherogenesis in general, not just in FH.

12.2. The Identity of the Scavenger Receptor's Substrate

The discovery of the scavenger receptor was a crucial step in the development of the inflammatory-response-to-injury theory of atherosclerosis, which was originally proposed by Virchow and has since been elaborated by Libby[3] and others. The nature of the physiologically relevant chemical modifications of LDL that made it a suitable substrate for the scavenger receptor of macrophages became a topic of intensive research by Dan Steinberg and his colleagues at the University of California, San Diego, in La Jolla. This began in 1981 when he and his co-authors reported that incubation of ^{125}I-labelled human LDL with rabbit aortic endothelial cells resulted in an increase in the density and electrophoretic mobility of the [125]I- LDL. They further showed that endothelial cell-modified LDL was taken up and degraded by macrophages 3–4 times faster than control LDL. The nature of the modification was not established in this particular study, but the authors suggested that it might play a significant role in atherogenesis.[4]

A breakthrough came in 1984 when Steinberg and his co-workers discovered that modification of LDL by rabbit aortic endothelial cells involved degradation of 40% of its phosphatidylcholine to lysophosphatidylcholine and that this and other sequelae of the modification process could be prevented by incubating the endothelial cells with an antioxidant such as vitamin E.[5] Changes in LDL similar to those resulting from incubation with endothelial cells were induced by exposure of LDL to an oxidising agent such as copper ions. The authors speculated that if lipid peroxidation of LDL occurs *in vivo*, then it might be possible to inhibit atherogenesis by treatment with antioxidants.

12.3. The Oxidised LDL Theory of Atherosclerosis

Although definitive proof of a causal role for oxidised LDL is lacking, there is ample evidence that it is present in atherosclerotic lesions where it stimulates the release of pro-inflammatory cytokines, which induce the formation of vascular cell adhesion molecules and promote the attachment of blood-borne monocytes to the endothelium. After migrating into the arterial intima, monocytes take on the characteristics of macrophages and express scavenger receptors. The ensuing uptake of modified LDL results in the formation of cholesterol ester engorged foam cells and leads to the formation of atheromatous plaques. The presence of severe hypercholesterolaemia from birth, as seen in FH homozygotes, accelerates this process and leads to the premature onset of aortic and coronary atherosclerosis and myocardial infarction. The evidence that oxidation of LDL is an essential feature of atherosclerosis is based largely on the detailed studies conducted by Dan Steinberg and his colleagues over a period of more than 30 years, as summarised in what follows.

12.3.1. *Macrophages and oxidised LDL*

In 1986, Parthasarathy *et al.*[6] showed that not only endothelial cells but also macrophages had the ability to induce oxidation of LDL, which was then taken up and degraded by the macrophages themselves via their scavenger receptors. A subsequent study confirmed that scavenger receptors had a high affinity for acetyl LDL but not for LDL itself and that uptake of endothelial cell-modified ^{125}I-LDL was competitively inhibited by acetyl LDL. The data also showed that the scavenger receptor recognised an epitope (antigenic determinant) on the apo B of LDL, not an oxidised lipid moiety,[7] although the latter did exert a strong chemotactic action on human monocyte-macrophages.

Steinberg and his colleagues therefore proposed that if elevated LDL levels were accompanied by cell-mediated oxidative modification of both the lipid and protein moieties of LDL, this could lead not only to the recruitment of macrophages but also to their subsequent transformation into lipid-laden foam cells, resulting in formation of fatty streaks, the earliest lesion of atherosclerosis.[8]

12.3.2. *Studies with probucol*

Studies with the lipid-lowering compound probucol showed that it prevented oxidative modification of LDL both by endothelial cells and by copper ions *in vitro*.[9] Furthermore, LDL isolated from probucol-treated subjects was resistant to oxidative modification, implying that this effect might reduce foam cell formation *in vivo*. Subsequent studies in genetically hypercholesterolaemic rabbits showed that probucol did indeed reduce the rate of development of fatty streaks in the aorta compared with statin-treated rabbits with similar cholesterol levels.[10] Paradoxically, however, despite its anti-oxidant properties probucol had a pro-atherogenic effect in LDL receptor deficient mice, which was attributed to its HDL lowering effect.[11] The latter property eventually led to the abandonment of probucol as a lipid-lowering compound in humans.

12.3.3. *Oxidised LDL and atherogenesis*

Incubation of LDL with endothelial cells is accompanied by degradation of its protein component apoB to smaller peptides. This process is secondary to peroxidation of LDL lipids and is completely prevented by the addition of anti-oxidants to the incubation medium.[12] Antibodies against oxidised LDL react with components of atherosclerotic lesions in LDL receptor-deficient rabbits[13] and additional evidence that atherosclerotic lesions contain oxidised LDL has come from the similarity of the LDL extracted from human and rabbit lesions with LDL oxidised *in vitro*.[14] However, despite the strength of the evidence that LDL undergoes oxidation and that oxidised LDL is found in atherosclerotic lesions, it remains unknown as to how and where this process occurs *in vivo*.[15] In a review article in 2001 Joe Witztum and Steinberg summarised the data supporting the presence of oxidised LDL in animal models and the many experimental studies that showed that inhibiting LDL oxidation retarded atherogenesis, but stressed the need to design appropriate clinical trials to test the oxidised LDL hypothesis in humans.[16]

12.3.4. *Clinical trials of anti-oxidants*

In 2009, Dan Steinberg reviewed the accumulated data, much of it from his own laboratory, that supported the oxidative modification of LDL hypothesis of atherosclerosis.[17] The evidence from *in vitro* experiments and studies in animal models is strongly supportive of the role of oxidised LDL. So too are the data that the severity of atherosclerosis in several animal models can be ameliorated by treatment with a variety of anti-oxidants, notably probucol and vitamin E. Further evidence that oxidised LDL is atherogenic came from a study showing that knocking out the gene for paroxonase-1, an enzyme that inhibits LDL oxidation, enhanced atherosclerotic lesion formation in apoE-deficient mice by 50%.

The following year Steinberg and Witzum published an update summarising the complexities of the oxidised LDL hypothesis of atherosclerosis.[18] They started by pointing out that the reason why endothelial cell-modified LDL was denser and more negatively charged than native LDL was due to the hydrolysis of oxidised phospholipids and the masking of lysine groups on LDL-apoB. Since these changes were blocked by vitamin E *in vitro*, Steinberg and Witzum concluded that they resulted from oxidation of LDL. Furthermore, as stated above, the anti-oxidant probucol inhibited atherosclerosis in genetically hypercholesterolaemic rabbits, supporting the atherogenic role of oxidised LDL *in vivo*. Finally, the presence of oxidised LDL in experimental animals and humans was confirmed by monoclonal antibodies that reacted with oxidation-specific epitopes on LDL.

However, this convincing trail of evidence came to an abrupt halt when the oxidised LDL hypothesis was tested in clinical trials. A meta-analysis of data from 80,000 persons showed no overall benefit from the administration of vitamin E on cardiovascular outcomes.[19] The meta-analysis included 15 controlled clinical trials, each comprising ≥1,000 post-myocardial infarction or high-risk subjects who were followed up for one to 12 years. Anti-oxidant treatment consisted of natural or synthetic vitamin E administered in doses of 400 i.u. or 300 mg daily. respectively, or β-carotene 20–50 mg daily, or a cocktail of vitamin E and β-carotene with or without vitamin C 250 mg daily. A total of 10 trials showed no reduction in cardiovascular morbidity or mortality, two showed reductions in non-fatal myocardial infarction and three showed adverse effects on cardiovascular endpoints. The authors of the report, including Steinberg and Witzum, concluded that the scientific data available at the time did not justify the use of anti-oxidant vitamin supplements for CVD risk reduction.

The main avenue of support for the relevance of the oxidised LDL hypothesis to human disease has come from studies conducted in persons under oxidative stress, such as those with end-stage renal disease on haemodialysis and diabetics with the haptoglobin 2-2 genotype. In both these groups there were highly significant reductions in cardiovascular events in patients treated with vitamin E. Future studies are required to determine whether there is an optimal dose of vitamin E or some other anti-oxidant needed to achieve such an effect in the general population, and if so, what should be the timing of this intervention in relation to the stage of development of atherosclerotic lesions.

For the time being it is premature to dismiss the role of oxidised LDL in atherosclerosis with Thomas Huxley's quotation "A beautiful hypothesis slayed by an ugly fact." Instead, acknowledging Dan Steinberg's Californian affiliations, it seems more appropriate to use Bob Dylan's contemporary expression "the answer is blowin' in the wind." Hopefully the winds of science will blow in the right direction and provide the solution to this intriguing and important conundrum.

12.4. Biography

Dan Steinberg was born in 1922. He obtained an MD degree at Wayne State University in Detroit in 1944 and went on to get a PhD in Biological Chemistry at Harvard in 1950. For the following 17 years he worked at the NHLBI in Bethesda where he investigated the role of hormone-sensitive lipase in regulating fat metabolism and also discovered the metabolic defect responsible for Refsum's Disease, namely an inherited deficiency of an enzyme that causes accumulation of phytanic acid. In 1968, he moved to the University of California, San Diego where he remained for the rest of his career.

While there he was a principal investigator of one of the centres involved in the Lipid Research Clinics Coronary Primary Prevention Trial and also headed one of the first Specialised Centers of Research (SCOR) funded by the NIH. However, as detailed in this chapter, his main focus of research while at La Jolla was on the oxidised LDL hypothesis of atherosclerosis (Fig. 12.1).

In 1984, he chaired the NIH Development Conference on Lowering Blood Cholesterol to prevent Coronary Heart Disease, which markedly influenced subsequent public health policy in the USA. The various experts who participated were asked to decide whether the relationship between blood cholesterol and coronary heart disease was causal, whether reduction of blood cholesterol would prevent coronary heart disease and, if so, whether an attempt should be made to reduce cholesterol levels in the general population.

The panel concluded that the answer to each of these questions was yes and it recommended treating all Americans adults and children with cholesterol levels above the 75th percentile with lipid-lowering measures, initially consisting of dietary advice. It further concluded that the NHLBI

Fig. 12.1.   Dan Steinberg (1922–2015) (2nd from the left in the front row) flanked by Bob Levy and Tony Gotto in 1975.

should develop plans for a National Cholesterol Education Program (NCEP) and it encouraged physicians to measure their patients' cholesterol levels on an opportunistic basis. Lastly, the report stressed the need for basic, clinical, epidemiological and pharmacological research into the role of cholesterol in atherosclerosis.[20]

Steinberg had been elected to the National Academy of Sciences 2 years earlier and received numerous awards and honours in his lifetime including the Lucian Award from McGill University in 1987. He was the Levi Professor of Medicine and Ageing at UC San Diego until 2000 when he became a Research Professor Emeritus. In 2007, he published a widely acclaimed book, *The Cholesterol Wars: the Skeptics vs. the preponderance of evidence.*[21]

In one of the last papers he published, he commemorated the 100th anniversary of Anitschkow's description of atherosclerosis in cholesterol-fed rabbits and reiterated the historic role it played in helping to establish the Lipid Hypothesis.[22] However, the question of whether oxidised LDL is an integral feature of atherogenesis was left unanswered during Dan's lifetime and remains so to this day.

He died in 2015 aged 92 and is greatly missed by all who had the privilege of knowing him, a charming and cultured man. The author fondly remembers the occasion when he and his wife were invited by Dan and his wife to take afternoon tea with them at Claridges Hotel during one of their frequent visits to London. He may have been small in stature, but he was a giant intellectually.

References

1. Goldstein JL, Ho YK, Basu SK, *et al*. Binding site on macrophages that mediates uptake and degradation of acetylated low density lipoprotein, producing massive cholesterol deposition. *Proc. Natl. Acad. Sci. USA* 1979; **76**: 333–337.
2. Brown MS, Goldstein JL, Krieger M, *et al*. Reversible accumulation of cholesteryl esters in macrophages incubated with acetylated lipoproteins. *J. Cell Biol.* 1979; **82**: 597–613.
3. Libby P. Inflammation in atherosclerosis. *Nature* 2002; **420**: 868–874.
4. Henriksen T, Mahoney EM, Steinberg D. Enhanced macrophage degradation of low density lipoprotein previously incubated with cultured endothelial cells: Recognition by receptors for acetylated low density lipoproteins. *Proc. Natl. Acad. Sci.* 1981; **78**: 6499–6650.
5. Steinbrecher UP, Parthasarathy S, Leake DS, Witztum JL, Steinberg D. Modification of low density lipoprotein by endothelial cells involves lipid peroxidation and degradation of low density lipoprotein phospholipids. *Proc. Natl. Acad. Sci. USA* 1984; **81**: 3883–3887.
6. Parthasarathy S, Printz DJ, Boyd D, Joy L, Steinberg D. Macrophage oxidation of low density lipoprotein generates a modified form recognized by the scavenger receptor. *Arteriosclerosis* 1986; **6**: 505–510.
7. Parthasarathy S, Fong LG, Otero D, Steinberg D. Recognition of solubilized apoproteins from delipidated, oxidized low density lipoprotein (LDL) by the acetyl-LDL receptor. *Proc. Natl. Acad. Sci. USA* 1987; **84**: 537–540.
8. Quinn MT, Parthasarathy S, Fong LG, Steinberg D. Oxidatively modified low density lipoproteins: A potential role in recruitment and retention of monocyte/macrophages during atherogenesis. *Proc. Natl. Acad. Sci. USA* 1987; **84**: 2995–2998.
9. Parthasarathy S, Young SG, Witztum JL, Pittman RC, Steinberg D. Probucol inhibits oxidative modification of low density lipoprotein. *J. Clin. Invest.* 1986; **77**: 641–644.
10. Carew TE, Schwenke DC, Steinberg D. Antiatherogenic effect of probucol unrelated to its hypocholesterolemic effect: Evidence that antioxidants in vivo can selectively inhibit low density lipoprotein degradation in

macrophage-rich fatty streaks and slow the progression of atherosclerosis in the Watanabe heritable hyperlipidemic rabbit. *Proc. Natl. Acad. Sci. USA* 1987; **84**: 7725–7729.

11. Bird DA, Tangirala RK, Fruebis J, *et al*. Effect of probucol on LDL oxidation and atherosclerosis in LDL receptor-deficient mice. *J. Lipid Res.* 1998; **39**: 1079–1090.

12. Fong LG, Parthasarathy S, Witztum JL, Steinberg D. Nonenzymatic oxidative cleavage of peptide bonds in apoprotein B-100. *J. Lipid Res.* 1987; **28**: 1466–1477.

13. Palinski W, Rosenfeld ME, Ylä-Herttuala S, *et al*. Low density lipoprotein undergoes oxidative modification in vivo. *Proc. Natl. Acad. Sci. USA* 1989; **86**: 1372–1376.

14. Ylä-Herttuala S, Palinski W, Rosenfeld ME, *et al*. Evidence for the presence of oxidatively modified low density lipoprotein in atherosclerotic lesions of rabbit and man. *J. Clin. Invest.* 1989; **84**: 1086–1095.

15. Chisolm GM, Steinberg D. The oxidative modification hypothesis of atherogenesis: An overview. *Free Radic. Biol. Med.* 2000; **28**: 1815–1826.

16. Witztum JL, Steinberg D. The oxidative modification hypothesis of atherosclerosis: Does it hold for humans? *Trends Cardiovasc. Med.* 2001; **11**: 93–102.

17. Steinberg D. The LDL modification hypothesis of atherogenesis: An update. *J. Lipid Res.* 2009; **50**(Suppl): S376–81.

18. Steinberg D, Witztum JL. Oxidized low-density lipoprotein and atherosclerosis. *Arterioscler. Thromb. Vasc. Biol.* 2010; **12**: 2311–2316.

19. Kris-Etherton PM, Lichtenstein AH, Howard BV, Steinberg D, Witzum JL. Antioxidant vitamin supplements and cardiovascular disease. *Circulation* 2004; **110**: 637–641.

20. Lowering blood cholesterol to prevent heart disease. NIH consensus development conference statement. *Arteriosclerosis* 1985; **5**: 404–412.

21. Glass CK, Witzum JL. Daniel Steinberg, 1922–2015. *Proc. Natl. Acad. Sci. USA* 2015; **112**: 9791–9792.

22. Steinberg D. In celebration of the 100th anniversary of the lipid hypothesis of atherosclerosis. *J. Lipid Res.* 2013; **54**: 2946–2949.

Chapter 13

Kåre Berg: The Discovery
of Lipoprotein (a)

13.1. History of Lp(a)

The Norwegian geneticist Kåre Berg discovered lipoprotein (a) (Lp(a)) in 1963.[1] Using an anti-serum from rabbits immunised with β lipoprotein (LDL), he found that one-third of individuals in the population expressed a previously unknown antigen that shared epitopes with LDL but contained an additional antigen. At the time he called this antigen Lp(+), which later became known as Lp(a), and he continued to undertake research on the latter for the remainder of the 20th century. In 1975, he and his co-workers used agarose gel electrophoresis to show a strong, positive association between the presence in serum of pre-beta1-lipoprotein (a lipoprotein with preβ mobility present in the d > 1.006 fraction of plasma, also known as "sinking" preβ lipoprotein) and the presence of Lp(+) in healthy controls and patients with a previous myocardial infarction (MI). Both markers were more prevalent in the MI survivors than in the controls[2] and they appeared to be identical to Lp(a).[3]

A subsequent study showed that Lp(a) levels were genetically determined and demonstrated a significant effect of a single, autosomal locus on a lipid implicated in atherosclerosis.[4] This finding was confirmed by an angiographic study which showed that the severity of coronary atherosclerosis was associated both with the presence of Lp(a) and a family history of coronary heart disease.[5] In 1987, Berg cited recent studies that indicated a population-attributable risk of 28% for MI in men below the

139

age of 60 with Lp(a) levels in the top quartile and he concluded that a raised level of Lp(a) was a major genetic risk factor for premature CHD.[6]

In 1990, Ikuko Kondo and Berg showed that the antigenicity of Lp(a) resided in a polypeptide chain attached to the apolipoprotein B (apoB) component of Lp(a) by a disulphide bridge (a covalent link). The cDNA for this polypeptide chain, known as apolipoprotein (a), was cloned and extensive homology with plasminogen was revealed.[7] This resulted in speculations that Lp(a) might exert an inhibitory effect on fibrinolysis, but there was no evidence of this in an *in vitro* study.[8] Despite this negative result, Berg considered that Lp(a) might have thrombogenic as well as atherogenic properties and he proposed that it should be measured in people with premature coronary heart disease or a positive family history.[9]

In 1998, Berg and his compatriot Terje Pedersen analysed the results of 4S, in which hypercholesterolaemic patients with coronary disease were randomised to receive simvastatin or a placebo.[10] Baseline Lp(a) levels in 4S patients were strikingly higher than in healthy controls. Furthermore, the number of deaths in the simvastatin group differed significantly between quartiles of Lp(a) lipoprotein levels, the reduction in deaths being most pronounced in those in the next to lowest quartile of Lp(a). Subjects with major coronary events had significantly higher Lp(a) levels than subjects without such events, re-affirming that a raised level of Lp(a) is a significant risk factor for coronary heart disease.[11]

In 2000, in a search for factors influencing Lp(a) levels, Oddveig Røsby and Berg[12] investigated the size polymorphism of apo(a) and showed that Lp(a) levels were inversely correlated with isoforms (similar but non-identical proteins encoded by the same gene) having the lowest number of Kringle IV repeats in apo(a). In the same year, Berg collaborated with Santica Marcovina and others to show that the size heterogeneity of apo(a) adversely affected the accuracy of isoform-sensitive immunochemical methods used to quantify Lp(a),[13] which underlined the need to develop isoform-insensitive assays. In the final 5 years of his research into Lp(a) Berg moved with the times and switched to using transgenic mice expressing the human LPA gene, having advanced knowledge of Lp(a) in humans as far as he could through his clinical and genetic studies.

13.2. Biography

Kåre Berg was born in the northeast of Norway in 1932 but grew up in the southeast of the country. He graduated with a degree in Medicine from the University of Oslo in 1957 and in 1960, after internships and military service, he became resident in internal medicine at Ulleval Hospital, Oslo. From 1964–1967 he was a research fellow at Rockefeller University, New York. On his return to Norway he was appointed Professor of Medical Genetics in Oslo University and in 1976 he became head of the department of Medical Genetics at Ulleval Hospital. This involved providing genetic counselling for the city of Oslo and he retained this role until he retired in 2002. He was an advisor in genetic disease and medical ethics to the WHO, editor in chief of *Clinical Genetics* from 1970 to 1997 and in 1994 he was elected to the Norwegian Academy of Science and Letters.

In 1992, the author was invited to contribute to a Festschrift organised and published by Oslo University for Kåre Berg's 60th birthday (Fig. 13.1).[14] The Festschrift had as its theme the Genetics of Coronary

Fig. 13.1.   Kåre Berg (1932–2009) (Photo: Øivind Larsen).

Heart Disease with a focus on Lp(a), the genetic risk factor Berg discovered almost 30 years earlier, but the clinical significance of which is only now becoming fully appreciated. He died in 2009 aged 76.

13.3. Contemporary Knowledge of Lp(a)

Contemporary knowledge of the structure and metabolism of Lp(a) is based to a considerable extent upon the pioneering discoveries of Berg and his colleagues over the years and there is inevitably a certain amount of repetition of facts in the following review of the current status of Lp(a).

13.3.1. *Structure and metabolism of Lp(a)*

Lp(a) consists of an LDL particle to which is attached a glycoprotein, apolipoprotein (a) or apo(a), covalently linked to the apolipoprotein B (apoB) moiety of LDL by a disulphide bond. The presence of apo(a) increases the density of Lp(a) compared with LDL and greatly reduces its affinity for the LDL receptor. This presumably explains why raised Lp(a) levels in plasma are unaffected by statins, which increase hepatic LDL receptor activity and enhance the catabolism of LDL but not of Lp(a).

Apo(a) consists of a number of pleated structures, so-called Kringles, one of which, Kringle IV type 2, is repeated a variable number of times, from 2 to >40; this gives rise to considerable differences in the size and molecular weight of apo(a) between individuals, the number of repeats being determined by the size of the apo(a) gene.[15] Lp(a) levels are thus largely genetically determined, being unaltered by other CVD risk factors or by life style.

Variation in the size and molecular weight of apo(a) between individuals has important implications for the accurate measurement of Lp(a). Apo(a) isoform size and Lp(a) concentration are inversely correlated; hence inheritance of small isoforms, with few Kringle IV type 2 repeats, results in higher Lp(a) levels than inheritance of larger isoforms, reflecting differences in hepatic synthesis rates.

The plasma concentration of Lp(a) ranges from 0 to >2,000 mg/l and is markedly skewed towards lower values in European populations, 500 mg/l being the 80th percentile.[16] Turnover studies suggest that the assembly of Lp(a) from newly synthesised LDL and apo(a) takes place intracellularly in the liver[17] and the increased production of LDL apoB evident in familial hypercholesterolaemia (FH) may explain why FH patients have Lp(a) levels that are two to threefold higher than normal subjects matched for isoform size.[15] With regard to catabolism, there is evidence that the kidney plays an important role in the catabolism and excretion of Lp(a).

13.3.2. *Status of Lp(a) as a risk factor*

Three studies published in 2009 provide strong support for the role of Lp(a) as a cardiovascular risk factor.[16] The largest was the Emerging Risk Factors Collaboration meta-analysis of prospective studies, which showed that Lp(a) levels in the range 500–2,000 mg/l were independently associated with a 15–25% increase in risk of non-fatal myocardial infarction and coronary death, together with a smaller increase in the risk of ischaemic stroke.[18] The causal nature of this association was supported by two Mendelian randomisation studies which showed a doubling of the risk of myocardial infarction[19] and coronary disease[20] at Lp(a) levels of 1,000 mg/l. Variations in the Lp(a) gene were the strongest correlate of risk in these studies.

In 2010, a European Atherosclerosis Society Consensus Panel advised that Lp(a) should be measured at least once in subjects known to be at high risk of CVD, in patients with FH, and in those with statin-refractory dyslipidaemia and CVD. Values >500 mg/l are considered to be undesirably high[16] and were recently shown to improve prediction of the risk of myocardial infarction, particularly in individuals at intermediate risk on conventional criteria.[21] The association between raised Lp(a) and CVD events persists even in subjects whose LDL cholesterol has been lowered to <1.8 mmol/l by statin therapy.[22]

13.3.3. *Assay of Lp(a)*

The variable number of Kringle IV type 2 repeats in apo(a) poses methodological problems for the immunoassay of Lp(a) in plasma. Marcovina *et al.*[23] showed that a monoclonal antibody that did not recognise Kringle IV type 2 repeats gave identical results to those obtained with a polyclonal antibody directed against the apoB component of Lp(a). In contrast, an isoform-sensitive monoclonal antibody directed against the variable Kringle IV type 2 region of apo(a) overestimated Lp(a) concentrations in samples with >21 Kringle IV type 2 repeats and underestimated Lp(a) concentrations in samples with <21 repeats. Since the lower the number of repeats the higher the concentration of Lp(a), isoform-sensitive assays tend to underestimate pathologically high levels.

Comparative studies of 19 different assays showed that the one with the best concordance with the gold standard isoform-insensitive method was an immuno-turbidometric assay developed by Denka Seiken.[13] Since all such assays quantify the protein component of Lp(a), the latter's concentration is now expressed as nmol/l of Lp(a) protein rather than as total mass (to roughly convert mg/dl of mass to nmol/l of protein, multiply by 2.3).

A major limitation of the widely-used Friedewald equation for calculating LDL cholesterol is that it includes the cholesterol carried in Lp(a). Direct estimation of the proportion of the total mass of Lp(a) that is present as cholesterol gave a mean value of 25%.[24] On this basis a person with an Lp(a) concentration of 100 mg/dl will have an Lp(a) cholesterol concentration of 25 mg/dl or 0.65 mmol/l. Assuming that this individual has an uncorrected Friedewald- estimated LDL cholesterol of 2.5 mmol/l, the actual value of LDL cholesterol will be 1.85 mmol/l. This is highly relevant if the individual concerned is on statin therapy and their clinician is frustrated by his or her apparent inability to achieve a target level of < 2 mmol/l. Hence, apparent statin-resistance may simply be an indication of a raised Lp(a).

13.3.4. *Treatment of raised Lp(a)*

Hitherto the best evidence that lowering Lp(a) reduces risk has come from studies using lipoprotein apheresis. However, this procedure must be performed weekly on a long-term basis and represents a major commitment by patient, physician and funding bodies. Selective Lp(a) apheresis, using a specific immunoadsorptive column, is a useful research tool but apheresis techniques which remove both LDL and Lp(a) are commonly used in clinical practice.

The status of Lp(a) as a risk factor is unequivocal and developing a safe and effective pharmacological means of lowering Lp(a) remains a high priority. In the meantime, increased efforts should be made to optimise treatment of concomitant CVD risk factors, especially a raised LDL, while the use of antiplatelet or antithrombotic agents may mitigate the putative thrombogenicity of Lp(a).

13.3.5. *Evidence of therapeutic benefit from lowering Lp(a)*

Since an elevated Lp(a) level is an independent risk factor for atherosclerosis, it is to be expected that lowering Lp(a) levels will translate into clinical benefit. It is however unclear how much Lp (a) must be decreased to achieve a significant reduction in risk. A recent study based on genetic data hypothesized that a decrease in Lp(a) concentration of >100 mg/dl is required to achieve a benefit equivalent to 1 mmol/l (39 mg/dl) of LDL-cholesterol lowering.[25] On the other hand, data from the Odyssey Outcomes trial of a PCSK9 inhibitor[26] indicate that a lesser reduction in Lp(a) was beneficial (1 mg/dl reduction resulted in 0.6% relative risk reduction — thus, about 35 mg/dl lipoprotein (a) reduction would lead to the same risk reduction as 1 mmol/L (39 mg/dl) of LDL-C reduction).

This topic is further complicated by the fact that as yet there are no drugs currently licensed that solely decrease Lp(a) concentrations. PCSK9-inhibitors can decrease Lp(a) concentrations but primarily decrease LDL-cholesterol, which makes it difficult to decide how much of the clinical benefit relates to LDL-cholesterol reduction and how much to lipoprotein (a) reduction. Similarly, most lipoprotein apheresis methods decrease both LDL and Lp(a) concentrations, again making it difficult to dissect out the effect of Lp(a) reduction.

An analysis of the German Lipoprotein Apheresis Registry for the period 2012–2015 showed acute reductions of LDL-cholesterol and Lp(a) of 69% and 70%, respectively.[27] The data showed a dramatic reduction (–97%) in cardiovascular events when the period before initiation of apheresis was compared to the period after regular apheresis. This analysis confirms previous German studies and one from Italy that evaluated cardiovascular events before the initiation of and during regular apheresis therapy.[28-31] In two of the German studies only subjects with isolated Lp(a) elevation were included (with LDL cholesterol <2.5 mmol/l (97 mg/dl) on statin therapy),[30,31] while in the third patients with concomitantly elevated LDL-cholesterol were also included.[29] The event rate decreased in all four studies dramatically after initiation of regular apheresis, but interpretation of these observations is hampered by the lack of any control groups.[32,33]

The US National Lipid Association and Heart UK guidelines consider elevated Lp(a) as an additional risk factor that should be taken into account when deciding whether lipoprotein apheresis should be used to treat elevated LDL-C, but an elevated Lp(a) per se is not regarded as an indication.[24] In contrast, elevated Lp(a) levels are considered an indication for regular apheresis in Germany if certain prerequisites are fulfilled, namely if Lp(a) is >60 mg/dl in patients with progressive cardiovascular disease despite optimal management of all other risk factors including LDL-C.[27]

13.4. Advances in Lp(a)-Lowering Drugs

Potent compounds for lowering Lp(a) are now being developed and tested. Phase 2 clinical trials of pelacarsen have shown that this antisense oligonucleotide to apo(a) reduced mean Lp(a) levels by 80%, enabling 98% of subjects to reach on-treatment levels of <125 nmol/l (~50 mg dl).[34] The phase 3 Lp(a)HORIZON outcomes trial is currently enrolling approximately 7,680 patients with a raised Lp(a) and a history of myocardial infarction, ischemic stroke and symptomatic peripheral arterial disease and a controlled LDL cholesterol, who will be randomised to pelacarsen or placebo. The co-primary endpoints are major adverse cardiovascular events in subjects with Lp(a) >70 mg/dl and >90 mg/dl. Additional RNA-targeted therapies to lower Lp(a) are in preclinical and clinical development. If successful in reducing cardiovascular events, these pharmacological approaches to lowering Lp(a) will render the use of apheresis for this purpose redundant, but in the interim it fulfils a useful function.

13.5. Conclusions

The importance of Berg's discovery of Lp(a) is obvious in retrospect but was initially overlooked by those working in the field of lipids and atherosclerosis. This was because, being a geneticist, Berg initially published his findings in genetic journals. His work provided insights into the structure, genetics and epidemiology of Lp(a), although not its physiological function, which remains obscure. The role of Lp(a) as a causal risk factor for atherosclerosis needs to be confirmed by selectively lowering raised plasma levels of Lp(a) and reversing that risk, a question that should soon be resolved by the HORIZON trial. Assuming that the outcome is positive, Berg will rightfully belong to the select group of scientists who have made a major contribution to the scientific basis of the lipid hypothesis.

References

1. Berg K. A new serum type system in man: The Lp system. *Acta Pathol. Microbiol. Scand.* 1963; **59**: 369–382.
2. Dahlén G, Berg K, Gillnäs T, Ericson C. Lp(a) lipoprotein/pre-beta1-lipoprotein in Swedish middle-aged males and in patients with coronary heart disease. *Clin. Genet.* 1975; **7**: 334–341.
3. Dahlén G, Frick MH, Berg K, Valle M, Wiljasalo M. Further studies of Lp(a) lipoprotein/pre- beta1-lipoprotein in patients with coronary heart disease. *Clin. Genet.* 1975; **8**: 183–189.
4. Berg K, Hames C, Dahlén G, Frick MH, Krishan I. Genetic lipoprotein variation and lipid levels in man. *Clin. Genet.* 1976; **10**: 97–103.
5. Frick MH, Dahlén G, Berg K, Valle M, Hekali P. Serum lipids in angiographically assessed coronary atherosclerosis. *Chest* 1978; **73**: 62–65.
6. Berg K. Genetics of coronary heart disease and its risk factors. *Ciba Found Symp.* 1987; **130**: 14–33.
7. Kondo I, Berg K. Inherited quantitative DNA variation in the LPA ("apolipoprotein (a)") gene. *Clin. Genet.* 1990; **37**: 132–140.
8. Halvorsen S, Skjønsberg OH, Berg K, Ruyter R, Godal HC. Does Lp(a) lipoprotein inhibit the fibrinolytic system? *Thromb. Res.* 1992; **68**: 223–232.
9. Berg K. Lp(a) lipoprotein: An overview. *Chem. Phys. Lipids* 1994; **67–68**: 9–16.
10. Pedersen TR, Olsson AG, Faergeman O, *et al.* Lipoprotein changes and reduction in the incidence of major coronary heart disease events in the Scandinavian Simvastatin Survival Study (4S). 1998. *Atheroscler. Suppl.* 2004; **5**: 99–106.
11. Berg K, Dahlén G, Christophersen B, *et al.* Lp(a) lipoprotein level predicts survival and major coronary events in the Scandinavian Simvastatin Survival Study. *Clin. Genet.* 1997; **52**: 254–261.
12. Røsby O, Berg K. LPA gene: Interaction between the apolipoprotein (a) size ("kringle IV" repeat) polymorphism and a pentanucleotide repeat polymorphism influences Lp(a) lipoprotein level. *J. Intern. Med.* 2000; **247**: 139–152.
13. Marcovina SM, Albers JJ, Scanu AM *et al.* Use of a reference material proposed by the International Federation of Clinical Chemistry and Laboratory Medicine to evaluate analytical methods for the determination of plasma lipoprotein (a). *Clin. Chem.* 2000; **46**: 1956–1967.
14. Thompson GR. Strategies for treating patients with high levels of Lp(a) lipoprotein. In: *Genetics of Coronary Heart Disease.* Bearn AG (ed.) Institute of medical genetics, University of Oslo; 1992.
15. Kronenberg F, Utermann G. Lipoprotein (a): Resurrected by genetics. *J. Intern. Med.* 2013; **273**: 6–30.

16. Nordestgaard BG, Chapman MJ, Ray K, *et al.* Lipoprotein (a) as a cardio-vascular risk factor: Current status. *Eur. Ht. J.* 2010; **31**: 2844–2853.
17. Frishman ME, Ikewaki K, Trenkwalder E, *et al.* In vivo stable-isotope study suggests intracellular assembly of lipoprotein (a). *Atherosclerosis* 2012; **225**: 322–327.
18. Erqou S, Kaptoge S, Perry PL, *et al.* Lipoprotein (a) concentration and the risk of coronary heart disease, stroke, and nonvascular mortality. *JAMA* 2009; **302**: 412–423.
19. Kamstrup PR, Tybjaerg-Hansen A, Steffensen R, *et al.* Genetically elevated lipoprotein (a) and increased risk of myocardial infarction. *JAMA* 2009; **301**: 2331–2339.
20. Clarke R, Peden JF, Hopewell JC, *et al.* Genetic variants associated with Lp(a) lipoprotein level and coronary disease. *N. Engl. J. Med.* 2009; **361**: 2518–2528.
21. Kamstrup PR, Tybjaerg-Hansen A, Nordestgaard BG. Extreme lipopro-tein (a) levels and improved cardiovascular risk prediction. *J. Am. Coll. Cardiol.* 2013; **61**: 1146–1156.
22. Albers JJ, Slee A, O'Brien KD, *et al.*, Relationship of apolipoproteins A-1 and B, and lipoprotein (a) to cardiovascular outcomes in the AIM-HIGH trial. *J. Am. Coll. Cardiol.* 2013; **62**: 1575–1579.
23. Marcovina SM, Albers JJ, Gabel B, *et al.* Effect of the number of apopro-tein (a) Kringle 4 domains on immunochemical measurements of lipopro-tein (a). *Clin. Chem.* 1995; **41**: 246–255.
24. Kulkarni KR, Garber DW, Marcovina SM, *et al.* Quantification of choles-terol in all lipoprotein classes by the VAP-II method. *J. Lipid Res.* 1994; **35**: 159–168.
25. Burgess S, Ference BA, Staley JR, Freitag DF, Mason AM, Nielsen SF, *et al.* Association of LPA variants with risk of coronary disease and the implica-tions for lipoprotein (a)-lowering therapies: A Mendelian randomization analysis. *JAMA Cardiol.* 2018; **3**: 619–627.
26. O'Donoghue ML, Fazio S, Giugliano RP, *et al.* Lipoprotein (a), PCSK9 inhibition, and cardiovascular risk. *Circulation* 2019; **139**: 1483–1492.
27. Schettler VJ, Neumann CL, Peter C, Zimmermann T, *et al.* The German Lipoprotein Apheresis Registry (GLAR) — Almost five years on. *Clin. Res. Cardiol. Suppl.* 2017; **12**(Suppl 1): 44–49.
28. Bigazzi F, Sbrana F, Berretti D, Maria Grazia Z, Zambon S, Fabris A, *et al.* Reduced incidence of cardiovascular events in hyper-Lp(a) patients on lipoprotein apheresis. The G.I.L.A. (Gruppo Interdisciplinare Aferesi Lipoproteica) pilot study. *Transfus. Apher. Sci.* 2018; **57**: 661–664.
29. Jaeger BR, Richter Y, Nagel D, Heigl F, Vogt A, Roeseler E, *et al.* Longitudinal cohort study on the effectiveness of lipid apheresis treatment to reduce high lipoprotein (a) levels and prevent major adverse coronary events. *Nat. Clin. Pract. Cardiovasc. Med.* 2009; **6**: 229–239.

30. Leebmann J, Roeseler E, Julius U, Heigl F, Spitthoever R, Heutling D, *et al.* Lipoprotein apheresis in patients with maximally tolerated lipid-lowering therapy, lipoprotein (a)-hyperlipoproteinemia, and progressive cardiovascular disease: prospective observational multicenter study. *Circulation* 2013; **128**: 2567–2376.

31. Rosada A, Kassner U, Vogt A, Willhauck M, Parhofer K, Steinhagen-Thiessen E. Does regular lipid apheresis in patients with isolated elevated lipoprotein (a) levels reduce the incidence of cardiovascular events? *Artif. Organs* 2014; **38**: 135–141.

32. Waldmann E, Parhofer KG. Lipoprotein apheresis to treat elevated lipoprotein (a). *J. Lipid Res.* 2016; **57**: 1751–1757.

33. Waldmann E, Parhofer KG. Apheresis for severe hypercholesterolaemia and elevated lipoprotein (a). *Pathology* 2019; **51**: 227–232.

34. Tsimikas S, Moriarty PM, Stroes ES. Emerging RNA therapeutics to lower blood levels of Lp(a): JACC focus seminar 2/4. *J. Am. Coll. Cardiol.* 2021; **77**: 1576–1589.

Chapter 14

Gerd Utermann: Discovery of ApoE Polymorphism and Its Role in Lipid Metabolism

14.1. Introduction

The discovery of apoE polymorphism by Gerd Utermann has had wide-ranging consequences. It not only revealed the cause of type III hyper-lipoproteinaemia but also showed that the apoE phenotype influences serum cholesterol and triglyceride levels in the population at large. Inherited differences in the frequency of the three common apoE alleles help explain racial differences in lipid levels and predict the likelihood of developing Alzheimer's disease. Also, a recently described novel deletion in apoE is a rare cause of familial hypercholesterolaemia.

14.2. ApoE2 and Type III Hyperlipoproteinaemia

The clinical and laboratory features of type III hyperlipoproteinaemia (dysβlipoproteinaemia), namely equimolar increases in serum cholesterol and triglyceride, a broad β-migrating band on agarose gel electrophoresis of the $d < 1.006$ fraction of plasma (β-VLDL), tubero-eruptive xanthomas, xanthelasma, palmar striae and an increased risk of vascular disease were described by Fredrickson $et\ al.$ in 1967.[1] The cause was unknown at the time, but those authors speculated that the disorder represented the homozygous expression of an uncommon mutant gene that resulted in an abnormal lipoprotein with the density of VLDL but with the electrophoretic mobility of LDL, hence the term dysβlipoproteinaemia. It was shown subsequently that the so-called abnormal lipoprotein represented the accumulation in plasma of chylomicron and VLDL remnant particles. Twelve years earlier, Gofman *et al.* had described marked increases in S_f 12–400 lipoproteins in patients with tuberose xanthomas, who almost certainly had the same disorder (see Chapter 3).

In 1973, Havel and Kane[2] described increased amounts of an arginine-rich protein in remnant particles from dysbetalipoproteinaemic subjects after a fatty meal. Almost simultaneously Shore and Shore[3] showed that the arginine-rich protein existed in three polymorphic forms. The breakthrough came in 1975 when Utermann isolated from VLDL and characterised the arginine-rich apoprotein, which he named apoE.[4] In that same year he and his colleagues reported that VLDL from normal subjects contained three arginine-rich polypeptides which they termed apoE-I, II and III. The third of these polypeptides was absent from the remnant particles of type III patients and Utermann *et al.* hypothesised that a recessively inherited deficiency of this isoform of apoE was the cause of the disorder.[5]

Further light was thrown on the matter in 1977 by Utermann *et al.* in a paper in *Nature*.[6] Using isoelectric focussing they showed that apoE polymorphism was determined by two alleles which they termed apoE^n and apoE^d. Patients with type III hyperlipoproteinaemia were homozygous for apoE^d. So too was 1% of the German population although only a small minority of this 1% were hyperlipidaemic. The authors suggested that additional genetic factors had to coexist with apoE^d homozygosity to produce the full-blown type III phenotype. They also described a third form of apoE that occurred in 27% of the samples analysed, which they termed apo E-IV (+).

Work by later investigators, using 2-dimensional polyacrylamide gel electrophoresis, resulted in a conflicting nomenclature for the various apoE isoforms. To resolve the confusion the various groups involved got together and agreed upon a common nomenclature reflecting the existence of three major alleles for apoE, giving rise to 6 commonly occurring apoE phenotypes.[7] Their approximate prevalences are E2/2 (0.4%), E2/3 (6.5%), E2/4 (0.9%), E3/3 (75.9%), E3/4 (14.3%) and E4/4 (2.0%), respectively.[8] E3/E3 is by far the commonest phenotype and apoE3 is regarded as the wild type or "normal" isoform of apoE.

Subsequent research by Mahley *et al.*[9] showed that apoE2 differs from apoE3 and apoE4 by the substitution of cysteine for arginine at amino acid 158. This causes a change in the conformation of apoE2 that inhibits its ability to act as a ligand for the LDL (apoB/E) receptor. The latter binds and mediates the uptake of VLDL remnants containing predominantly apoE3 and apoE4 but not those containing apoE2, which explains the accumulation of remnant particles in type III subjects. Most patients with type III hyperlipoproteinaemia are homozygous for apoE2, but some are heterozygotes or have inherited one of the 31 rare variants of apoE described in a recent review.[8] Some of these variants give rise to a dominant expression of the disorder whereas with others it is recessive.

The commonest cause of type III (apoE2/apoE2) behaves in a phenotypically recessive manner, namely it expresses itself as hyperlipidaemia only if additional risk factors are present. The latter occur in only a minority of instances (1:5,000) and include familial hypercholesterolaemia and hypothyroidism, both of which reduce the number of LDL receptors available to bind remnants. In addition, factors which enhance secretion of chylomicrons and VLDL, such as a high fat diet, excess alcohol, obesity and non-insulin-dependent diabetes increase the accumulation of remnant particles in the plasma of subjects with functionally defective apoE isoforms.

Dominant forms of type III result from mutations that cause irreversible changes in the receptor-binding domain of apoE. These include amino acid substitutions at positions 136 (serine for arginine), 145 (cysteine for arginine) and 146 (glutamine for lysine). The 145 (cysteine for arginine) mutation is particularly prevalent among the black community in the vicinity of Cape Town.[10] These three variants all have apoE2 mobility on electrophoresis whereas apoE$_{Leiden}$, which results from a seven amino acid insertion but has apoE3 electrophoretic mobility, also causes dominant expression of type III. ApoE phenotyping has now been replaced by apoE genotyping as the preferred means of diagnosing type III in the lipid clinic.

14.3. ApoE Polymorphism in Various Populations

In 1987, Utermann reviewed the role of apoE polymorphism in influencing lipid levels in populations worldwide.[11] He found that the ε2 allele was commoner in Chinese, Indians and Malays than in Germans and Finns whereas ε4 was commonest in Finns and least common in the Japanese. Within the German and Finnish populations serum cholesterol levels were highest in in those with an ε4 allele and lowest in those with an *ℰ*2 allele. Differences in cholesterol were attributed on the one hand to the subnormal binding (compared with apoE3) of apoE2-containing lipoproteins to hepatic LDL receptors, the consequent upregulation of the latter resulting in lowering of LDL. In contrast, apoE4-containing particles are catabolised more rapidly in vivo than apoE three-containing lipoproteins, with consequent down regulation of hepatic LDL receptors and an increase in LDL, especially in populations consuming a high cholesterol diet such as the Finns.

In subjects with various forms of primary hyperlipidaemia, apoE2 heterozygosity is associated with hypertriglyceridaemia and the *ℰ*4 allele with hypercholesterolaemia, while both the *ℰ*2 and *ℰ*4 alleles occur more frequently in mixed hyperlipidaemic than in normal subjects.[11] A survey of over 500 German blood donors by Boerwinkle and Utermann showed that compared with those with an *ℰ*3 allele, possession of an *ℰ*2 allele lowered apoB and total cholesterol levels while having an *ℰ*4 allele did the opposite. Overall apoE polymorphism accounted for 12% of the variability of apoB and 4% of the variability of total cholesterol in these presumably healthy subjects.[12]

ApoE polymorphism also influences the response to treatment of hyperlipidaemic subjects. In a study of statin efficacy in patients with familial hypercholesterolaemia, those with an *ℰ*4 allele responded poorly to statin therapy, possibly because hepatic cholesterol synthesis was already downregulated by apoE4-enhanced cholesterol absorption. In contrast, good responders to statins had fewer apoE4 alleles and higher pre-treatment rates of cholesterol synthesis, presumably reflecting upregulation of HMG CoA reductase secondary to decreased cholesterol absorption.[13] These examples illustrate the significant role played by apoE polymorphism in influencing LDL receptor expression in the liver, both in normal subjects and in those with hyperlipidaemia.

14.4. ApoE4 and Alzheimer's Disease

Alzheimer's disease, the commonest cause of dementia, currently affects about 1 million people in the UK and is projected to rise to 1.7 million in 2051.[14] A meta-analysis of European population-based studies performed in the 1990s, the EURODEM study,[15] demonstrated the progressive increase in the incidence and prevalence of dementia with increasing age, especially in women. More than 95% of those with Alzheimer's disease, which causes 62% of cases of dementia, are over the age of 65.[14]

Risk factors for vascular dementia include diabetes, hypertension, smoking and obesity, but their role in Alzheimer's disease is unclear. In contrast, there is good evidence that genetic factors are involved. Three genes are implicated in Early-onset Alzheimer's disease, amyloid precursor protein (*APP*) and two presenilin genes (*PSEN* 1 and *PSEN* 2). Mutations in any of these genes can result in an imbalance between amyloid-β peptide (Aβ) production and clearance that leads to a build-up of Aβ.

The likelihood of developing Late-onset Alzheimer's disease is markedly influenced by the apoE genotype, which accounts for 20–30% of the risk. More than 55% of the UK population are homozygous for ε3, roughly 25% are heterozygous for ε3 and ε4, approximately 3% are homozygous for ε4 and most of the remainder are heterozygous for ε2.[16] Compared with individuals homozygous for ε3, possession of one ε4 allele increases the risk of Late-onset Alzheimer's disease by 2 to 4-fold whereas homozygosity for ε4 increases the risk almost 15-fold. Life-time risks of developing Alzheimer's disease by age 85 are 5–6% in ε2/2+2/3, 7–10% in ε3/3, 20–30% in ε2/4 and 3/4, and 50–60% in ε4/4; the higher values in each instance relate to females.[17] Other genes implicated in Late-onset Alzheimer's disease include Sortilin-related receptor 1 (*SORL* 1) and clusterin (*CLU*). The latter, also known as lipoprotein J, forms complexes with Aβ in the cerebro-spinal fluid that can penetrate the blood-brain barrier.

14.4.1. *Pathogenesis of Alzheimer's disease*

The twin pathological hallmarks of Alzheimer's disease are the accumulation in the cerebral cortex of plaques consisting of extracellular deposits of Aβ peptide, and intracellular accumulation of neurofibrillary tangles containing hyperphosphorylated tau (*p*-tau), a microtubule assembly protein.[18] These changes are accompanied by loss of neurons in the temporal lobes and a progressive decline in cognitive function. Mechanisms invoked for the increased risk associated with the ε4 allele include interaction between apoE4 and Aβ, and apoE4-mediated inhibition of neuronal repair and remodelling. In vitro incubation of Aβ with apoE4 results in the formation of a dense, stable network of neurofibrils, whereas incubation with apoE3 results in a less dense, less stable network. The more complex plaques of Aβ formed in the presence of apoE4 may impair the normal clearance mechanism and thereby enhance plaque formation and/or promote the formation of neuro-fibrillary tangles.

Thus, apoE3 appears to facilitate cytoskeletal activity in the cerebral cortex while apoE4 may inhibit it.[19] An isoform-dependent effect on cholesterol efflux, ApoE2 > ApoE3 > ApoE4, has been demonstrated in primary rat or mouse astrocytes and neurons.[20] This ranking correlates inversely with the likelihood of humans carrying these apoE alleles developing AD,[17] but it remains to be established whether impaired efflux of cholesterol contributes to the pathogenesis of the disorder in man.

14.4.2. *ApoE polymorphism and cardiovascular disease*

The influence of apoE polymorphism on CHD risk in 14 published studies was analysed by Wilson *et al.*,[21] who found that possession of an ε4 allele was associated with a 26% increase in risk compared with individuals homozygous for ε3. Another more recent study involving over 10,000 control subjects found that the frequency of ε4 homozygotes decreased from 2.7% in those below the age of 60–0.8% in those aged over 85, the corresponding values in individuals with an ε3/4 genotype being 26.8% and 17.5%, respectively.[22] Increased mortality from CHD and Alzheimer's disease seems the most likely explanation for the faster decrease in the frequency of ε4 carriers with increasing age relative to those with ε2 and ε3 alleles.

Interaction between the apoE genotype and statin therapy was examined in a sub-study of the Scandinavian Simvastatin Survival Study (4S).[23] This showed that the risk of death among myocardial infarct survivors with an ε4 allele was more than twice that of non-carriers and that this excess risk was abolished by treatment with simvastatin. The latter observation implies that by offsetting the increased cardiovascular risk and thereby increasing their life expectancy, statins might increase the chances of ε4 carriers developing Alzheimer's disease — unless they also mitigate the risk of the latter.

14.4.3. *Statins and dementia*

Case control[24] and prospective[25] studies have suggested that statins decrease the risk of Alzheimer's disease by around 40%. The Rotterdam Study found that this protective effect was exerted irrespective of whether the statins used were lipophilic and crossed the blood-brain barrier or were hydrophilic and did not, which is counterintuitive. In contrast to these observational studies, an updated Cochrane review of two double-blind randomised, placebo-controlled trials in people at risk of dementia, the Prospective Study of Pravastatin in the Elderly at Risk (PROSPER) and the Heart Protection Study (HPS), found no evidence that statins, whether hydrophilic or lipophilic, can prevent dementia.[26] A more recent analysis of PROSPER showed that administering pravastatin to men and women aged 70–82 years had no effect on the decline in cognitive function with age and it was concluded that giving statins to the elderly to prevent cognitive decline is a futile exercise.[27]

A universally effective treatment for Alzheimer's disease remains a highly desirable but elusive goal at present. However, a very recent report describes two missense variants, one coinherited with apoE3 and the other with apoE4, that are associated with a decreased risk of Alzheimer's disease.[28] The location of these variants confirms that the carboxy-terminal portion of apoE plays an important role in the pathogenesis of this disorder. Understanding the mechanism of the protective effect could well lead to the development of drugs that might be beneficial in Alzheimer's disease.

14.5. ApoE and Familial Hypercholesterolaemia

In the great majority of patients with autosomal dominant familial hyper-cholesterolaemia (FH) the underlying defect is a pathogenic mutation of the LDL receptor gene or, in a minority of instances, mutations of the apoB and PCSK9 genes. In 2013, Marduel *et al.*[29] described a large family in France with FH without any of these causes but with a novel mutation in the apoE gene, p. Leu167del. Kinetic studies in an affected carrier demonstrated increased production and decreased catabolism of LDL. Later that year Awan *et al.*[30] described a patient in Canada with barn door FH (LDL cholesterol 9.7 mmol/l, tendon xanthomas, xanthelasma and myocardial infarct at age 43) who had the same mutation, an in-frame three base-pair deletion that affected the receptor binding domain of apoE.

The prevalence of apoE Leu167del in 229 French children and adults with FH was 1.7%[31] and was 3.1% in 288 Spanish patients with FH.[32] The latter study found that supra-normal amounts of mutant apoE in VLDL resulted in downregulation of LDL receptors and increased levels of LDL cholesterol. A subsequent *in silico* investigation of the mechanism involved suggested that strong binding of apoE *p.*Leu167del to the LDL (apoB/E) receptor impaired re-cycling of the latter, thereby reducing LDL catabolism.[33]

Recently, another Spanish study compared the response to intensive statin therapy of 22 FH patients with the apoE Leu167del mutation with that of 44 FH patients with an LDL receptor mutation.[34] The results showed a more marked decrease in LDL cholesterol in the former than in the latter, 52% vs. 40%, indicating that FH due to apoE Leu167del should be easier to treat than other causes.

These various reports demonstrate the role of apoE in mediating binding and uptake of apoB-containing lipoproteins and illustrate that mutations of the apoE gene, depending upon their nature and location, most often cause type III hyperlipoproteinaemia. However, they very occasionally cause FH, and genetic screening strategies for the latter should take this into account.[30]

14.6. Biography

Gerd Utermann was born in Germany in 1939, just after the start of the Second World War. He went to high school in Dortmund and then studied Medicine at the Universities of Marburg, Freiburg and Vienna, qualifying MD in 1969. He was a Research Associate in the Department of Biochemistry from 1968–1972 and in the Department of Human Genetics at Marburg between 1973–1984; he served his internships and residency there during the latter period. In 1984, he was appointed Professor of Human Genetics at the University of Marburg, but later that year he moved to the University of Innsbruck as Professor of Medical Biology and Human Genetics and was Head of the Institute of Medical Biology and Human Genetics until 2004. The following year he was appointed Director of the Department of Medical Genetics, Molecular and Clinical Pharmacology and from 2009 onwards, Emeritus Professor at the Institute of Human Genetics, Medical University of Innsbruck.

His hobbies include trout fishing, a pastime in which he and the author participated together on a Hampshire chalk stream many years ago. He is also a keen bird-watcher and was mortified while on a visit to the Kruger Park when despite his best efforts he failed to spot a Secretary Bird, whereas the author happened to see one purely by chance! He belies the stereotypic image of a German and has an excellent sense of humour (Fig. 14.1).

Fig. 14.1.   Gerd Utermann (grey T-shirt) with Roger Illingworth (1945–2013) (no shirt) and the author on top of Lion's Head with Table Mountain, Cape Town in the background (Photo: David Marais, 1990).

Gerd Utermann's contributions to research on apoE were recognised in 1985 when, against stiff opposition, he was chosen by an international committee to be the first winner of the G.B. Morgagni Medal of the University of Padua. The award was based on his scientific reputation, the originality of his research and its importance to mankind. This was only 10 years after he and his colleagues first isolated, characterised and named apoE, but the impact of their findings in advancing knowledge of the pathophysiology of lipids soon became apparent to those in the field, whether basic scientist or clinician. For example, the vital role of apoE3 or apoE4, but not apoE2, in enabling apoB/E receptor-mediated binding of VLDL remnants is vividly illustrated by subjects with familial defective apoB, in whom there is normal uptake of apoE3 or E4-containing remnants but not of apoB-containing LDL. In contrast, a profound defect in clearance of remnants but normal clearance of LDL characterises type III hyperlipoproteinaemia (apoE2/E2). Like FH, type III is another of Nature's hints, but in this instance the abnormality affects a ligand for the LDL receptor rather than the receptor itself.

References

1. Fredrickson DS, Levy RI, Lees RS. Fat transport in lipoproteins — An integrated approach to mechanisms and disorders. *N. Engl. J. Med.* 1967; **276**: 215–225.
2. Havel RJ, Kane JP. Primary dysbetalipoproteinemia: Predominance of a specific apoprotein species in triglyceride-rich lipoproteins. *Proc. Nat. Acad. Sci. USA* 1973; **70**: 2015–2019.
3. Shore VG, Shore B. Heterogeneity of human plasma very low density lipoproteins. Separation of species differing in protein components. *Biochemistry* 1973; **12**: 502–507.
4. Utermann G. Isolation and partial characterization of an arginine-rich apolipoprotein from human plasma very-low-density lipoproteins: Apolipoprotein E. *Hoppe Seylers Z Physiol. Chem.* 1975; **356**: 1113–1121.
5. Utermann G, Jaeschke M, Menzel J. Familial hyperlipoproteinemia type III: Deficiency of a specific apolipoprotein (apoE-III) in the very-low-density lipoproteins. *FEBS Lett.* 1975; **56**: 352–355.
6. Utermann G, Hees M, Steinmetz A. Polymorphism of apolipoprotein E and occurrence of dysbetalipoproteinaemia in man. *Nature* 1977; **269**: 604–607.
7. Zannis VI, Breslow JL, Utermann G, *et al.* Proposed nomenclature of apoE isoproteins, apoE genotypes, and phenotypes. *J. Lipid Res.* 1982; **23**: 911–914.
8. Khalil YA, Ràbes J-P, Boileau C, Varret M. *APOE* gene variants in primary dyslipidemia. *Atherosclerosis* 2021; **328**: 11–22.
9. Mahley RW, Innerarity TL, Rall SC, Weisgraber KH. Plasma lipoproteins: Apolipoprotein structure and function. *J. Lipid Res.* 1984; **25**: 1277–1294.
10. Blom DJ, Byrnes P, Jones S, Marais AD. Dysbetalipoproteinaemia — clinical and pathophysiological features. *S. Afr. Med. J.* 2002; **92**: 892–897.
11. Utermann G. Apolipoprotein E polymorphism in health and disease. *Am. Heart J.* 1987; **113**: 433–440.
12. Boerwinkle E, Utermann G. Simultaneous effects of the apolipoprotein E polymorphism on apolipoprotein E, apolipoprotein B, and cholesterol metabolism. *Am. J. Hum. Genet.* 1988; **42**: 104–112.
13. O'Neill FH, Patel DD, Knight BL, *et al.* Determinants of variable response to statin treatment in patients with refractory familial hypercholesterolaemia. *Arterioscler. Thromb. Vasc. Biol.* 2001; **21**: 832–837.

14. Alzheimer's Society. Statistics. http://alzheimers.org.uk/site/scripts/document_pdf.php?documentID=341.
15. Rocca WA, Hofman A, Brayne C, *et al.* Frequency and distribution of Alzheimer's disease in Europe: A collaborative study of 1980–1990 prevalence findings. The EURODEM-Prevalence Research Group. *Ann Neurol.* 1991; **30**: 381–390.
16. Wu K, Bowman R, Welch AA, *et al.* Apolipoprotein E polymorphisms, dietary fat and fibre, and serum lipids: The EPIC Norfolk study. *Eur. Heart J.* 2007; **28**: 2930–2936.
17. Genin E, Hannequin D, Wallon D, *et al. APOE* and Alzheimer disease: A major gene with semi-dominant inheritance. *Mol. Psy.* 2011; **16**: 903–907.
18. Reitz C, Brayne C, Mayeux R. Epidemiology of Alzheimer disease. *Nat. Rev. Neurol.* 2011; **7**: 137–152.
19. Mahley RW, Nathan BP, Pitas RE. Apolipoprotein E. Structure, function, and possible roles in Alzheimer's disease. *Ann. NY Acad. Sci.* 1996; **777**: 139–145.
20. Feringa FM, van der Kant R. Cholesterol and Alzheimer's disease; From risk genes to pathological effects. *Front Aging Neurosci.* 2021 June 24; **13**: 690372. doi: 10.3389/fnagi.2021.690372.
21. Wilson PW, Schaefer EJ, Larson MG, Ordovas JM. Apolipoprotein E alleles and risk of coronary disease. *Arterioscler. Thromb. Vasc. Biol.* 1996; **16**: 1250–1255.
22. McKay GJ, Silvestri G, Chakravarthy U, Dasan S, Fritsche LG, Weber BH *et al.* Variations in apolipoprotein E frequency with age in a pooled analysis of a large group of older people. *Am. J. Epidemiol.* 2011; **173**: 1357–1364.
23. Gerdes LU, Gerdes C, Kervinen K, Savolainen M, Klausen IC, Hansen PS, *et al.* The apolipoprotein *E*4 allele determines prognosis and the effect on prognosis of simvastatin in survivors of myocardial infarction. *Circulation* 2000; **101**: 1366–1371.
24. Zamrini E, McGwin G, Roseman JM. Association between statin use and Alzheimer's disease. *Neuroepidemiology* 2004; **23**: 94–98.
25. Haag MD, Hofman A, Koudstaal PJ, Stricker BH, Breteler MM. Statins are associated with a reduced risk of Alzheimer disease regardless of lipophilicity. The Rotterdam Study. *J. Neurol. Neurosurg. Psychiatry* 2009; **80**: 13–17.
26. McGuiness B, Craig D, Bullock R, Passmore P. Statins for the prevention of dementia. *Cochrane Database Syst. Rev.* 2009 Apr 15; (2): CD003160. doi: 10.1002/14651858.CD003160.pub2.
27. Trompet S, van Vliet P, de Craen AJ, Jolles J, Buckley BM, Murphy MB *et al.* Pravastatin and cognitive function in the elderly. Results of the PROSPER study. *J. Neurol.* 2010; **257**: 85–90.
28. Le Guen Y, Belloy ME, Grenier-Boley B, *et al.* Association of rare missense variants V236E and R215G with risk of Alzheimer disease. *JAMA Neurol.* **79**: 652–663. doi: 10.1001/jamaneurol.2022.1166.

29. Marduel M, Ouguerram K, Serre V, *et al*. Description of a large family with autosomal dominant hypercholesterolemia associated with the ApoE p.Leu167del mutation. *Hum. Mutat.* 2013; **34**: 83–87.
30. Awan Z, Choi HY, Stitziel N, *et al*. ApoE p.Leu167del mutation in familial hypercholesterolemia. *Atherosclerosis* 2013; **231**: 218–222.
31. Wintjens R, Bozon D, Belabbas K, *et al*. Global molecular analysis and apoE mutations in a cohort of autosomal dominant hypercholesterolemia patients in France. *J. Lipid Res.* 2016; **57**: 482–491.
32. Cenarro A, Etxebarria A, de Castro-Orós I, *et al*. The p. Leu167del mutation in apoE gene causes autosomal dominant hypercholesterolemia by down-regulation of LDL receptor expression in hepatocytes. *J. Clin. Endocrinol. Metab.* 2016; **101**: 2113–2121.
33. Rashidi OM, H Nazar FA, Alama MN, Awan ZA. Interpreting the mechanism of apoE (p.Leu167del) mutation in the incidence of familial hypercholesterolemia: An in silico approach. *Open Cardiovasc. Med. J.* 2017; **11**: 84–93.
34. Bea AM, Lamiquiz-Moneo I, Marco-Benedi V, *et al*. Lipid-lowering response in subjects with the (p. Leu167del) mutation in the apoE gene. *Atherosclerosis* 2019; **282**: 143–147.

Chapter 15

Akira Endo: The Discovery of the First Statin*

15.1. Introduction

The selection process for the Nobel Prize aims to identify milestones of scientific discovery.[1] This being so, it is surprising that Akira Endo has not yet been awarded one because his discovery in 1976[2] of the first statin, compactin, represented a scientific advance that had major implications for the prevention and treatment of atherosclerotic cardiovascular disease. Several more statins have been discovered or synthesised since then and the remarkable ability of these compounds to inhibit hepatic cholesterol synthesis, upregulate LDL receptors and thus lower serum cholesterol has resulted in them becoming among the most commonly used drugs in the world.

15.2. Biography

Akira Endo was born in November 1933 in Akita Prefecture in the northeast of Japan. His war time childhood was a rather sombre affair but was enlivened by his grandfather who used to take him mushroom picking in the nearby hills, engendering a life-long interest in fungi. He excelled in Chemistry at high school and obtained a scholarship to Tohoku University,

---

*Edited from: Thompson G, Mabuchi H. Akira Endo: The Discovery of Statins. In: *Pioneers of Medicine without a Nobel Prize*, Thompson G (ed.). London: Imperial College Press; 2014.

Sendai, where he studied biotechnology. Immediately after graduating in 1957, he joined the Sankyo Company in Tokyo where his first project in the applied microbiology group was to develop an enzyme that could hydrolyse pectins contaminating wines and ciders. He succeeded in purifying a new pectinase and for this he was awarded his PhD by Tohoku University in 1966.

The commercial success of his pectinase project provided Endo with the opportunity to study in the United States and he opted to work on lipids at Harvard with Konrad Bloch, who had recently won the Nobel Prize for elucidating the cholesterol biosynthetic pathway. Unfortunately, Bloch did not have a vacancy, but nevertheless encouraged his interest in cholesterol metabolism. Instead, from 1966–1968, Endo did his postdoctoral studies in the Department of Molecular Biology at Albert Einstein College of Medicine, working with Lawrence Rothfield on an enzyme involved in the biosynthesis of bacterial lipopolysaccharide. While in New York, Endo was astonished by the high prevalence of coronary heart disease and hypercholesterolaemia in the USA, in marked contrast with post-war Japan. On his return to Tokyo he re-joined Sankyo and in 1971 started searching for a microbial inhibitor of cholesterol biosynthesis. Two years later he discovered compactin, the first inhibitor of hydroxyl methyl glutaryl co-enzyme A (HMG CoA) reductase, the rate-limiting enzyme of cholesterol synthesis, and in collaboration with Hiroshi Mabuchi he subsequently demonstrated its cholesterol-lowering properties in human subjects.

In 1978, Endo left Sankyo for an Associate Professorship in the School of Agriculture of Tokyo University of Agriculture and Technology where he continued his studies of fungal inhibitors of cholesterol synthesis. He became a Full Professor in 1986 and retired in 1997. Currently, he is a Distinguished Professor Emeritus at Tokyo University of Agriculture and Technology and Director of Biopharm Research Laboratories in Tokyo. He was awarded the Japan Prize in 2006, the Lasker-DeBakey Clinical Research Award in 2008 (Fig. 15.1) and the Gairdner International Award in 2017. The author came to know and to like Akira Endo during his visits to Japan and appreciated his innate modesty and quiet sense of humour.

15.3. Cholesterol Metabolism and Coronary Heart Disease in the 1960s

Endo's period of working in New York coincided with a renaissance in lipid research in the United States, sparked by Fredrickson *et al.* in 1967.[3]

Fig. 15.1.   Akira Endo, flanked by the author and Tony Gotto, in New York after he was awarded the Lasker-DeBakey Clinical Research Award in 2008.

The mid-1960s were also the time when serum cholesterol levels and CHD mortality rates reached their respective peaks in the USA.[4] It is not surprising therefore that Endo took a keen interest in cholesterol metabolism and that he recognised the therapeutic potential of pharmacological inhibition of cholesterol synthesis.

In 1964, Bloch and Lynen were awarded the Nobel Prize in Physiology or Medicine for delineating the multitude of chemical reactions involved in the conversion of acetate to cholesterol. Four years earlier Siperstein and Guest[5] had proposed that the site of feedback regulation by cholesterol of its own synthesis was the conversion of HMG CoA to mevalonic acid, a relatively early step on the pathway. Subsequently, Siperstein and Fagan concluded that this feedback involved downregulation of the enzyme responsible for converting HMG CoA to mevalonic acid, namely HMG CoA reductase.[6] All that remained to be done was for someone to discover a competitive inhibitor of this enzyme — which is precisely what Endo aimed to do when he returned to Japan.

15.4. The Discovery of Compactin

Endo has described elsewhere[7] how he and Masao Kuroda started their search for microbial HMG CoA reductase inhibitors after his return to Sankyo in 1971. This was based on the premise that certain microorganisms might produce such compounds as a weapon against competitors that were dependent for their growth on sterol or isoprenoid metabolites of mevalonate. Inhibiting HMG CoA reductase would prevent synthesis of those metabolites and therefore be lethal to rival organisms.

15.4.1. *In vitro studies*

In order to screen microbial cultures for their ability to inhibit HMG CoA reductase Endo devised a cell-free assay system that measured cholesterol synthesis from ^{14}C-labelled acetate. For the next 2 years he painstakingly screened roughly 6,000 microbial cultures and eventually discovered a strain of *Penicillium citrinum* with inhibitory activity. In July 1973, he isolated the active component from 300 l of culture medium, ending up with 259 mg of a crystalline compound that was initially named ML-236B but was later called compactin or mevastatin.

Compactin inhibited cholesterol synthesis from both acetate and from HMG CoA in nanomolar concentrations but had no effect when mevalonate was the substrate. This proved that it was a potent inhibitor of HMG CoA reductase,[2,8] with an affinity for the enzyme that was 10,000-fold greater than the latter's natural substrate, HMG CoA. Using the cultured human fibroblast system devised by Brown and Goldstein for their research on LDL receptors, Endo and colleagues[9] were able to show that compactin was a potent inhibitor of cholesterol synthesis in cells from both a normal subject and a patient with homozygous FH.

By an interesting coincidence, workers at Beecham's Laboratories in the UK[10] had independently isolated compactin from *Penicillium brevicompactum* in 1976 while searching for anti-fungal agents. They meticulously described the structure of compactin, but its hypocholesterolaemic property was not investigated at Beecham's until 4 years later[11] and was then not exploited, which proved to be a lost opportunity in commercial terms.

15.4.2. *In vivo studies*

Oral administration of single 20 mg/kg doses of compactin to rats acutely reduced their serum cholesterol levels by 20–30%.[2] However, long-term administration to mice and rats was ineffective, probably because the marked compensatory increase in HMG CoA reductase activity that this evoked in the livers of these rodents counteracted the inhibitory effect of compactin.[11,12] In contrast, compactin lowered serum cholesterol by 45% in dogs[13] and also in monkeys, especially reducing their LDL cholesterol.[14]

The first clinical studies with compactin were carried out in 1980 by Yamamoto and Sudo in collaboration with Endo.[15] They showed that compactin in doses of 50–150 mg/day reduced serum cholesterol by 27% on average in patients with heterozygous FH or combined hyperlipidaemia, most of the decrease being in LDL cholesterol. Higher doses were needed to lower cholesterol significantly in FH homozygotes, one of whom developed a myopathy which resolved when the dose of compactin was reduced from 500 to 200 mg daily.

The following year Mabuchi *et al.* reported the results of administering compactin in a dose of 30–60 mg daily to seven FH heterozygotes for 24 weeks, which caused a 22% decrease in serum cholesterol.[16] A subsequent study[17] showed that co-administration of the bile acid sequestrant cholestyramine and compactin to FH heterozygotes resulted in an unprecedented 53% reduction in LDL cholesterol after 12 weeks, compared with a 28% reduction when cholestyramine 12 g daily was given alone. Reductions in LDL cholesterol were attributed to stimulation of receptor-mediated LDL catabolism by these compounds, which clearly had an additive effect when given together.

In an accompanying editorial, Brown and Goldstein[18] noted the dramatic reduction in LDL achieved with compactin but cautioned that many hurdles needed to be overcome before it could be regarded as a "penicillin" for hypercholesterolaemia. The wisdom of this comment soon became apparent when Sankyo suddenly suspended their clinical trial programme, reputedly because of the perceived carcinogenicity of extremely high doses of compactin in dogs (>500 mg/kg/day), and consequently it was never licensed for use in man.

15.5. Other HMG CoA Reductase Inhibitors

Endo had left Sankyo in 1978 to go to Tokyo University of Agriculture and Technology where he continued his studies of fungal inhibitors of cholesterol synthesis. Sankyo had collaborated with the US drug company Merck several years previously by providing them with samples of compactin. The head of research at Merck at the time, Roy Vagelos, had a long-standing interest in HMG CoA reductase and he, like Endo, realised the therapeutic potential of inhibiting this enzyme. In 1980, his colleague Al Alberts and others at Merck discovered a second HMG CoA reductase inhibitor, which they isolated from *Aspergillus terreus*.[19] This was initially called mevinolin but was later renamed lovastatin. Endo had isolated the same compound simultaneously but independently from a different mould, naming it monacolin K.[20] He went on to discover several other HMG CoA reductase inhibitors over the next few years before retiring from the university in 1997.

Merck commenced the clinical development of lovastatin, but the suspension of all clinical studies on compactin by Sankyo put them in a quandary, especially since Sankyo would not disclose the reason. After much deliberation, as described in detail by Steinberg,[21] Merck followed suit and stopped all clinical work while they undertook additional safety studies in animals. The results were sufficiently reassuring for them to resume clinical trials with lovastatin in 1983, and 4 years later lovastatin became the first HMG CoA reductase inhibitor to be licensed by the FDA.[22]

During the interim Merck arranged for lovastatin to be made available to treat patients with refractory familial hypercholesterolaemia on a compassionate use patient basis. The ability to treat these severely hypercholesterolaemic patients effectively and safely with such a well-tolerated compound represented a major therapeutic advance, especially since its LDL-lowering effect of >30 % was additive to that of existing therapies such as cholestyramine, partial ileal bypass and apheresis. However, as reported by Yamamoto *et al.*, it was evident that FH homozygotes were less responsive than heterozygotes.

Because of patent constraints by Sankyo, lovastatin was never licensed in Britain though Merck maintained supplies for compassionate use until 1989, when its successor simvastatin became available.

Simvastatin, a semi-synthetic derivative of lovastatin, differs only in having an additional methyl group which prolongs its duration of action. Other statins soon followed including pravastatin in 1991, which was derived from compactin and developed by Sankyo; fluvastatin in 1994, which was the first synthetic HMG CoA reductase inhibitor; atorvastatin in 1997 and rosuvastatin in 2003, the most potent of these compounds and capable of reducing LDL by >50%; and lastly pitavastatin, which was licensed in the USA in 2009. Cerivastatin was licensed in 1998 but then withdrawn by its manufacturers on account of its myotoxicity. The latter is the only serious side effect of statins but, cerivastatin apart, is rare with an incidence of <0.1%.[22] The myositis is reversible when the statin is stopped but if left undiagnosed can lead to rhabdomyolysis, renal failure and sometimes death.

The discovery of statins by Endo eventually resulted in huge profits for the pharmaceutical industry, but it was not always plain sailing. In addition to early concerns over the suspected toxicity of compactin and the withdrawal of cerivastatin for safety reasons, atorvastatin was nearly scrapped by Parke-Davis on economic grounds. It is said that Roger Newton, who was deeply involved in its development, literally went down on his knees before the Board to plead for its continuation. The directors relented and it eventually became the best-selling drug in the world.

15.6. Conclusions

Norrby discusses in *Nobel Prizes and Life Sciences*[1] the role of serendipity and quotes Pasteur's dictum that "chance favours the prepared mind." He gives numerous examples where serendipitous discoveries led to a Nobel Prize, emphasising the importance of determining who first made the critical observation. With regard to the discovery of statins, no one disputes Endo's primacy but nor would they claim that it was serendipitous. He undoubtedly had a prepared mind but his discovery was certainly not based on chance — he deliberately set out to search for an inhibitor of the key step in the regulation of cholesterol synthesis, using his knowledge of microbiology and lipid biochemistry to pursue this objective.

Mark Twain wrote eloquently about the joys of discovering a new idea, which in Japanese evokes the word "Ichiban" or being first. In any language Akira Endo is one of Medicine's pioneers and he deserves due recognition for initiating one of the most important therapeutic advances of recent times. His discovery of the cholesterol-lowering properties of *Penicillium citrinum* was as much a first in the treatment of cardiovascular disease as was Alexander Fleming's discovery half a century earlier of the antibiotic properties of *Penicillium notatum,* a first in the treatment of bacterial infections. Fleming, together with Chain and Florey, was awarded the Nobel Prize in 1945 for discovering and developing penicillin. Endo "whose discovery of statins changed the world" according to Brown and Goldstein[23] unaccountably lacks this recognition.

References

1. Norrby E. *Nobel Prizes and Life Sciences.* Singapore: World Scientific Publishing; 2010.
2. Endo A, Kuroda M, Tsujita Y. ML-236A, ML-236B, and ML-236C, new inhibitors of cholesterologenesis produced by Penicillium Citrinum. *J. Antibiot.* 1976; **29**: 1346–1348.
3. Fredrickson DS, Levy RI, Lees RS. Fat transport in lipoproteins — An integrated approach to mechanisms and disorders. *N. Eng. J. Med.* 1967; **276**: 34–42, 94–103, 148–156, 215–225, 273–281.
4. Cooper R, Cutler J, Desvigne-Nickens P, *et al.* Trends and disparities in coronary heart disease, stroke, and other cardiovascular diseases in the United States. *Circulation* 2000; **102**: 3137–3147.
5. Siperstein MD, Guest MJ. Studies on the site of the feedback control of cholesterol synthesis. *J. Clin. Invest.* 1960; **39**: 642–652.
6. Siperstein MD, Fagan VM. Feedback control of mevalonate synthesis by dietary cholesterol. *J. Biol. Chem.* 1966; **241**: 602–609.
7. Endo A. The discovery and development of HMG-CoA reductase inhibitors. *J. Lipid Res.* 1992; **33**: 1569–1582.
8. Endo A, Kuroda M, Tanzawa K. Competitive inhibition of 3-hydroxy-3-methylglutaryl coenzyme A reductase by ML-236A and ML-236B, fungal metabolites, having hypocholesterolemic activity. *FEBS Lett.* 1976; **72**: 323–326.
9. Kaneko I, Hazama-Shimada Y, Endo A. Inhibitory effects on lipid metabolism in cultured cells of ML-236B, a potent inhibitor of 3-hydroxy-3-methylglutaryl-coenzyme-A reductase. *Eur. J. Biochem.* 1978; **87**: 313–321.
10. Brown AG, Smale TC, King TJ, Hasenkamp R, Thompson RH. Crystal and molecular structure of compactin, a new antifungal metabolite from Penicillium brevicompactum. *J Chem Soc Perkin* 1. 1976; **11**: 1165–70. PMID: 945291.
11. Fears R, Richards DH, Ferres H. The effect of compactin, a potent inhibitor of 3-hydroxy-3-methylglutaryl coenzyme-A reductase activity, on

cholesterologenesis and serum cholesterol levels in rats and chicks. *Atherosclerosis* 1980; **35**: 439–449.

12. Endo A, Tsujita Y, Kuroda M, *et al.* Effects of ML-236B on cholesterol metabolism in mice and rats: Lack of hypocholesterolemic activity in normal animals. *Biochim. Biophys. Acta* 1979; **575**: 266–276.

13. Tsujita Y, Kuroda M, Tanzawa K, *et al.* Hypolipidemic effects in dogs of ML-236B, a competitive inhibitor of 3-hydroxy-3-methylglutaryl coenzyme A reductase. *Atherosclerosis* 1979; **32**: 307–313.

14. Kuroda M, Tsujita Y, Tanzawa K, *et al.* Hypolipidemic effects in monkeys of ML-236B, a competitive inhibitor of 3-hydroxy-3-methylglutaryl coenzyme A reductase. *Lipids* 1979; **14**: 585–589.

15. Yamamoto A, Sudo H, Endo A. Therapeutic effects of ML-236B in primary hypercholesterolemia. *Atherosclerosis* 1980; **35**: 259–266.

16. Mabuchi H, Haba T, Tatami R, *et al.* Effect of an inhibitor of 3-hydroxy-3-methylglutaryl coenzyme A reductase on serum lipoproteins and ubiquinone-10-levels in patients with familial hypercholesterolemia. *N. Eng. J. Med.* 1981; **305**: 478–482.

17. Mabuchi H, Sakai T, Sakai Y, *et al.* Reduction of serum cholesterol in heterozygous patients with familial hypercholesterolemia. *N. Eng. J. Med.* 1983; **308**: 609–613.

18. Brown MS, Goldstein JL. Lowering plasma cholesterol by raising LDL receptors. *N. Eng. J. Med.* 1981; **305**: 515–517.

19. Alberts AW, Chen J, Kuron G, Hunt V, *et al.* Mevinolin: A highly potent competitive inhibitor of hydroxymethylglutaryl-coenzyme A reductase and a cholesterol-lowering agent. *Proc. Natl. Acad. Sci. USA* 1980; **77**: 3957–3961.

20. Endo A. Monacolin K, a new hypocholesterolemic agent produced by a *Monascus* species. *J. Antibiot.* 1979; **32**: 852–854.

21. Steinberg D. An interpretative history of the cholesterol controversy, part V: The discovery of the statins and the end of the controversy. *J. Lipid Res.* 2006; **47**: 1339–1351.

22. Tobert JA. Lovastatin and beyond: The history of the HMG-CoA reductase inhibitors. *Nat. Rev. Drug Discov.* 2003; **2**: 517–526.

23. Brown MS, Goldstein JL. A tribute to Akira Endo, discoverer of a "Penicillin" for cholesterol. *Atheroscler. Supp.* 2004; **5**: 13–16.

Chapter 16

P. Roy Vagelos, Alfred Alberts and Jonathan Tobert: The Development of Statins for Clinical Use

16.1. Introduction

Although Endo discovered the first statin, compactin, this was never licensed for use in man and it was largely due to the work of Roy Vagelos and his colleagues at the US pharmaceutical company Merck, Al Alberts and Jonathan Tobert, that subsequent statins were discovered, developed and subjected to the safety tests required by regulatory authorities for licensing for clinical use. Their endeavours made a major contribution to the treatment of hypercholesterolaemia and atherosclerosis and resulted in large profits for Merck.

16.2. Vagelos: Biography

As recounted in his autobiography,[1] P. Roy Vagelos was born in 1929 in New Jersey, the son of Greek immigrants. After attending Rahway High School he graduated at the University of Pennsylvania and then went on to study Medicine at Columbia University's College of Physicians and Surgeons in New York. During the holidays he went home to Rahway and worked part-time in the nearby Merck Laboratories. After qualifying as an MD in 1954 he did his internship in Medicine at the Massachusetts General Hospital in Boston, followed by a post-doctoral fellowship in

Earl Stadtman's laboratory at the NIH, where he became involved in research on the biochemistry of lipids.

In 1959, he took on Al Alberts as his research assistant and they worked together since then, except for a sabbatical year that Vagelos spent in Jacques Monod's laboratory in Paris. In 1966, he was head-hunted and appointed chairman of the Department of Biological Chemistry at Washington University Medical School, St. Louis where he remained until 1975. During his time there he developed an interest in cholesterol metabolism in relation to heart disease, stimulated by the work of Goldstein and Brown in Dallas. Having discovered the role of the acetyl-CoA carrier protein in fatty acid synthesis he recognised the importance of HMG CoA reductase in regulating cholesterol synthesis and the therapeutic potential of inhibiting this enzyme.

In 1975, he accepted an offer from Merck to become Research Director and then President of Merck Research Laboratories, which meant moving back to Rahway. He was accompanied there by Alberts and after settling in they decided to focus their research activities on finding an inhibitor of HMG CoA reductase, the rate-limiting enzyme on the cholesterol synthesis pathway. They then had to choose whether to attempt to synthesise a new compound for this purpose or to search for an existing natural inhibitor in soil microorganisms. Stimulated by reports of Endo's discovery of compactin they set off down the natural inhibitor pathway and in 1978 they struck oil, as detailed in the succeeding narrative relating to Alberts's discovery and development of lovastatin.

In 1985, Vagelos was appointed Chief Executive Officer of Merck and in 1994 he reached the mandatory retirement age of 65. He left Merck to become Chairman of the Board of Trustees of the University of Pennsylvania and then became the Chairman of the Board of Regeneron Pharmaceuticals until 2023, when he finally retired at the age of 93.

16.3. Alfred Alberts

Al Alberts discovered and characterised lovastatin, the first statin in the world to be approved for the treatment of hypercholesterolaemia. He also played a key role in discovering the chemical modification of lovastatin that resulted in the synthesis of simvastatin. Both compounds have played a major role in the treatment and prevention of atherosclerotic cardiovascular disease.

16.3.1. *The discovery of mevinolin (lovastatin)*

Following the discovery of compactin in 1976, Merck concluded a confidentiality agreement with Sankyo which provided Merck with samples of crystalline compactin and relevant biochemical, pharmacological and toxicological data. Over the next 2 years Alberts and his colleagues undertook a series of *in vitro* and *in vivo* animal studies with compactin which confirmed Endo's findings. This stimulated them to develop a rapid, high throughput assay for screening microorganisms for HMG CoA reductase-inhibitory activity. One such compound was discovered by them in 1980 in a soil sample containing *Aspergillus terreus*. Subsequent structural studies demonstrated that this compound was distinct from compactin and it was initially named mevinolin (Mevacor), which was later changed to lovastatin.[2]

Alberts *et al.* showed that lovastatin was a more potent inhibitor than compactin of sterol synthesis in cultured mouse fibroblasts. The drug was also shown to inhibit cholesterol synthesis in rats after oral administration as either the lactone or hydroxyl acid. However, chronic administration did not lower rats' serum cholesterol owing to a compensatory increase in hepatic HMG CoA reductase. In contrast, it lowered serum cholesterol in dogs by nearly 30%. Even greater decreases were observed in cholestyramine-treated dogs after co-administration of lovastatin, attributable to an additive increase in hepatic LDL receptor activity. Hepatic uptake of lovastatin was shown to be enhanced by administering it as the inactive lactone, which subsequently underwent conversion in the liver to the active hydroxyl acid.[3]

Merck commenced the clinical development of lovastatin, but the sudden suspension of clinical studies with compactin by Sankyo for undisclosed reasons put them in a quandary. In view of rumours that compactin was carcinogenic, Merck followed suit and stopped all clinical work on lovastatin while they undertook additional safety studies in animals. The results were sufficiently reassuring for them to resume clinical trials in 1983 and 4 years later lovastatin became the first HMG CoA reductase inhibitor to be licensed for clinical use.[4]

16.3.2. *The synthesis of simvastatin*

During the pause in the clinical development of lovastatin alluded to above, Alberts and his chemist colleagues at Merck undertook a systematic exploration of the structure–activity relationships of a series of side chain ester derivatives of lovastatin.[5] They found that the addition of a methyl group to the side chain of lovastatin produced a compound with more than twice its HMG CoA reductase-inhibitory activity; this was initially named synvinolin but is now known as simvastatin. Rats treated for 4 days with 0.02% lovastatin and simvastatin added to their diet showed decreases in serum cholesterol of 33% and 64%, respectively, with analogous differential effects of the two compounds *in vitro*. In humans a comparative eight-week study in hypercholesterolaemic subjects showed that doses of 40 mg daily of lovastatin and simvastatin lowered low density lipoprotein cholesterol levels by 30% and 40%, respectively, confirming the greater efficacy of simvastatin.

However, greater efficacy was not the only advantage of simvastatin from Merck's perspective. Although the mevinolin version of lovastatin had been discovered 3 months before Endo's monacolin K version,[6] the patent for the latter was registered first. This enabled Sankyo to block the marketing of lovastatin in countries which recognised "time of application" for a patent, such as the UK and most of Europe, in contrast to countries which recognised "time of invention" such as the US and Canada. Fortunately for Merck no such constraints applied to simvastatin and it earned the company billions of dollars while it remained under patent. The clinical development of lovastatin and simvastatin is described in subsequent sections of this chapter.

16.3.3. *Biography*

Alfred Alberts was born in New York on May 16th, 1931. After getting his Bachelor of Science degree in Biology at Brooklyn College, New York, in 1953, he served with the US Army Medical Corps for 2 years and was then a graduate student in Zoology at the University of Maryland from 1955–1960. Before finishing his PhD dissertation (he never did) he joined Earl Stadtman's biochemistry laboratory at the National Institutes of Health, Bethesda, where he worked with Roy Vagelos on microbial enzymes of fatty acid synthesis, forming a partnership with Vagelos that lasted for over 30 years. The two of them moved together to Washington University School of Medicine, St. Louis, after Vagelos was appointed Chairman of Biochemistry. In 1973, Alberts became an Associate Professor of Biological Chemistry there, but he moved again with Vagelos when the latter was appointed Head of Merck Research Laboratories. From 1975 onwards Alberts led the research at Merck which resulted in the discovery of lovastatin and the development of simvastatin.

Alberts became Director of Natural Product Discovery at Merck and remained there until his retirement in 1995. He was awarded an Honorary Doctor of Science degree by the University of Maryland in 1994. He was a likeable, dedicated but underestimated scientist who spoke with a New York accent one could cut with a knife. He died in 2018, aged 87.

16.4. Merck's Dilemma

In 1980, while attending Merck's Annual Conference and basking in the reflected glory from the discovery of lovastatin, Vagelos received a phone call from one of his research executives in Japan that caused a sudden chill to descend on the proceedings. The caller broke the news that Sankyo had suspended all clinical studies with compactin due, it was rumoured, to evidence that large doses of the compound were carcinogenic in animals. Vagelos immediately suspended all clinical trials of lovastatin, informed the FDA and intensified animal toxicity studies with the drug. Despite his repeated efforts he was totally unable to get any information from Sankyo as to why they had suspended the development of compactin.

The initial clinical studies of lovastatin had begun with a phase 1 trial in healthy subjects that demonstrated a 35–45% reduction in LDL cholesterol with doses of 25–50 mg twice daily. By the time this was published in 1982, Merck had imposed its embargo on further clinical studies. Later that year, however, Merck initiated a compassionate use programme that enabled lipidologists in selected centres in the USA, Europe and the UK to treat very high- risk patients with familial hypercholesterolaemia. The encouraging results in these patients enabled Merck to resume clinical development of lovastatin in 1983 and undertake the phase 2 and phase 3 trials required to confirm its efficacy and safety. A new drug application was filed with the FDA in November 1986, and the clinical data were presented at an FDA Advisory Committee meeting in February 1987. The Committee unanimously recommended approval of lovastatin, which was ratified by the FDA in August 1987 (Fig. 16.1). The cover of the Reader's Digest in December that year hit the nail on the head with its caption "Wonder Drug that Zaps Cholesterol" — Vagelos's cautious gamble had paid off!

Fig. 16.1.   The Merck trio celebrating the licensing of lovastatin (N803MK) by the FDA in 1987: Front row from left to right Al Alberts (1931–2018), Roy Vagelos and Jonathan Tobert (Photo: Eli Alberts).

16.5. Jonathan Tobert

Jonathan Tobert was the "midwife" of lovastatin and simvastatin, playing a vital role in the delivery of these microbial-derived compounds from their laboratory womb into the harsh world of clinical medicine.

16.5.1. *The clinical testing of lovastatin and simvastatin*

Tobert was the clinical pharmacologist at Merck who was responsible for devising, supervising and publishing the initial clinical studies of lovastatin.[7–9] He was in charge of the compassionate use programme with lovastatin and the author was grateful to him for supplies of the latter during the 1980s to treat FH patients.[10] Because lovastatin did not have worldwide patent protection, Merck developed simvastatin and Tobert was involved with its clinical development too, particularly with the cardiovascular outcome studies 4S and HPS (the Scandinavian Simvastatin Survival study and the Heart Protection Study), working with the principal investigators in Oslo and Oxford. These studies were crucial in convincing the medical community that lowering LDL cholesterol could substantially reduce the risk of myocardial infarction and stroke, and by doing so reduce all-cause mortality. Their results helped prove statins to be highly effective and safe drugs that subsequently revolutionised the treatment of hypercholesterolaemia and cardiovascular disease.

16.5.2. *Statin intolerance*

Tobert and others identified muscle toxicity as the most serious but fortunately rare side effect of lovastatin and simvastatin and showed that the risk was increased by a variety of drugs which inhibited the activity of the cytochrome P450 pathway responsible for metabolising both these statins in the liver. More recently he became concerned about the misperception that many patients could not tolerate statins in clinical practice, leading to their stopping treatment, and pointed out that symptoms of statin intolerance are frequently not reproducible under double-blind conditions.

The question of statin intolerance has been a recurring topic of sometimes misleading comment in the lay press over the last few years. In a recent analysis of all the published evidence from clinical trials, Collins *et al.* estimated that statins cause symptomatic intolerance, notably muscle pain, in about 50–100 patients per 10,000 treated for 5 years (0.5–1.0%).[11] However, placebo-controlled randomised trials showed that almost all the adverse symptoms attributed to statin therapy in clinical practice were not actually caused by it but represented misattribution.

A possible explanation for the latter is the nocebo effect, the inverse of the placebo effect, a phenomenon that refers to adverse events that reflect subjective expectations of harm from a drug, driven by factors such as warnings by clinicians concerning adverse effects when prescribing it and misinformation in the media about its dangers. According to Tobert and Newman, the nocebo effect is the most likely explanation for the high rate of muscle and other symptoms attributed to statins in observational studies and clinical practice, because in randomized controlled trials the frequency of these symptoms was similar in the statin and placebo groups.[12]

Self-diagnosed statin-intolerant patients usually tolerate statins under double-blind conditions, which suggests that genuine statin intolerance is much less common than is claimed. But it does exist and when it occurs is either due to loss-of-function mutations of genes that modulate the transport of statins from the blood into the liver or to the interaction of statins with other drugs that patients are taking which get metabolised by the same liver enzymes. Both circumstances result in high blood levels of statins and predispose to genuine side-effects such as myopathy. Tobert's work has been instrumental in helping clinicians to discriminate between real and misattributed statin myalgia and enabling them to persuade patients in the latter category who have stopped statins to resume taking them.[13]

16.5.3. *Biography*

Jonathan Tobert was born in London, UK, on December 10, 1945. He obtained his medical degree from the University of Cambridge and Middlesex Hospital, London, in 1970 and a PhD in pharmacology from the University of London in 1975. After a year's research fellowship in Biological Chemistry at Harvard Medical School, Boston, he joined the Clinical Pharmacology Department of Merck Research Laboratories in Rahway, New Jersey, in 1976. He started work on lovastatin in 1979 and led the multidisciplinary project team throughout the drug's development, being responsible for designing and implementing the clinical studies required to demonstrate its efficacy and safety.

From 1988 to 1991, Tobert directed the Department of Clinical Endocrinology and Metabolism at Merck, and in addition to lovastatin and simvastatin he was also responsible for the 5-α reductase inhibitor finasteride for benign prostatic hyperplasia and the bisphosphonate bone resorption inhibitor alendronate. In 1991, he moved to a scientific track position and focused on evaluating the safety of statins and their effects on cardiovascular risk.

After 27 years at Merck, Tobert retired in 2004 and started a consulting company, providing advice to pharmaceutical companies on clinical development issues. He also became affiliated with the Nuffield Department of Population Health at the University of Oxford and was a member of the writing group that produced an American Heart Association Scientific Statement on statin-associated adverse effects.[14]

16.6. Postscript

Although Endo discovered the first statin there was no guarantee that Sankyo would have gone ahead with developing other statins had it not been for Merck. Roy Vagelos's deft handling of the toxicity issue, Al Alberts's timely discovery of lovastatin and synthesis of simvastatin, and Jonathan Tobert's meticulous collection of the clinical data required by the FDA, all contributed to the setting up of a grand finale to the statin saga in the shape of the clinical outcome trials.

References

1. Vagelos R, Galambos L. *Medicine, Science and Merck*. New York: Cambridge University Press; 2004.
2. Alberts AW, Chen J, Kuron G, *et al*. Mevinolin, a highly potent competitive inhibitor of HMG-CoA reductase and cholesterol lowering agent. *Proc. Natl. Acad. Sci. USA* 1980; **77**: 3957–3961.
3. Chao YS, Chen JS, Hunt VM, Kuron GW, Karkas JD, Liou R, Alberts AW. Lowering of plasma cholesterol levels in animals by lovastatin and simvastatin. *Eur. J. Clin. Pharmacol.* 1991; **40**: S11–S14.
4. Tobert JA. Lovastatin and beyond: The history of the HMG-CoA reductase inhibitors. *Nat. Rev. Drug Discov.* 2003; **2**: 517–526.
5. Hoffman WF, Alberts AW, Anderson PS, Chen JS, Smith RL, Willard AK. 3-hydroxy-3-methylglutaryl-coenzyme A reductase inhibitors. 4. Side chain ester derivatives of mevinolin. *J. Med. Chem.* 1986; **29**: 849–852.
6. Endo A. Monacolin K. A new hypocholesterolemic agent produced by a *Monascus* species. *J. Antibiot.* 1979; **32**: 852–854.
7. Tobert JA, Bell GD, Birtwell J, *et al*. Cholesterol-lowering effect of mevinolin, an inhibitor of 3-hydroxy-3-methylglutaryl-coenzyme A reductase, in healthy volunteers. *J. Clin. Invest.* 1982; **69**: 913–919.
8. Hunninghake DB, Miller VT, Palmer RH, *et al*. for Lovastatin Study Group II. Therapeutic response to lovastatin (mevinolin) in nonfamilial hypercholesterolemia: A multicenter study. *JAMA* 1986; **256**: 2829–2834.
9. Lovastatin Study Group III. A multicenter comparison of lovastatin and cholestyramine therapy for severe primary hypercholesterolemia. *JAMA* 1988; **260**: 359–366.
10. Thompson GR, Ford J, Jenkinson M, Trayner I. Efficacy of mevinolin as adjuvant therapy for refractory familial hypercholesterolaemia. *Q. J. Med.* 1986; **60**: 803–811.
11. Collins R, Reith C, Emberson J, *et al*. Interpretation of the evidence for the efficacy and safety of statin therapy. *Lancet* 2016; **388**: 2532–2561.
12. Tobert JA, Newman CB. The nocebo effect in the context of statin intolerance. *J. Clin. Lipidol.* 2016; **10**: 739–747.
13. Howard JP, Wood FA, Finegold JA, *et al*. Side effect patterns in a crossover trial of statin, placebo and no treatment. *J. Am. Coll. Cardiol.* 2021; **78**: 1210–1222.
14. Newman CB, Preiss D, Tobert JA, *et al*. Statin safety and associated adverse events: A scientific statement from the American Heart Association. *Arterioscler. Thromb. Vasc. Biol.* 2019; **39**: e38–e81.

Chapter 17

Terje Pedersen, James Shepherd and Sir Rory Collins: The Statin Trials that Proved the Lipid Hypothesis

17.1. Introduction

Despite the encouraging results of the Lipid Research Clinics Coronary Primary Prevention Trial (LRC–CPPT) and the NHLBI Type II Coronary Intervention Study, both published in 1984, many cardiologists of that era remained unconvinced that lowering cholesterol by diet or drugs would provide an effective and safe means of treating and preventing coronary heart disease. This view was epitomised by the adversarial attitude of McMichael and Mitchell, but there were some cardiologists, mainly American, who accepted the premise that cholesterol might play a causal role in atherosclerosis. They tested this hypothesis by undertaking a series of so-called regression trials, in which the impact of cholesterol-lowering measures on coronary artery disease in hyperlipidaemic patients was assessed by coronary angiography.

The chief instigator of these trials was the late David Blankenhorn from Los Angeles, generally regarded as the founding father of studies of atherosclerosis regression in humans (Fig. 17.1). Following on from his earlier studies of the effect of lipid-lowering therapy on femoral atherosclerosis, the Monitored Atherosclerosis Regression Study (MARS) was the first trial to demonstrate the effect of statin monotherapy in inducing angiographic regression of coronary atheroma.[1]

Fig. 17.1.   Joint meeting on "Regression" of British and European Atherosclerosis Societies at Downing College Cambridge in 1992. Numbered participants: 1 M Oliver, 2 M Davies, 3 G Brown, 4 D Blankenhorn, 5 G Shaper, 6 E Stein (chairman EAS), 7 T Mead (chairman BAS). The organisers Alan Howard and the author (brown suit) are at each end of the sofa.

17.2. Angiographic Trials of Lipid-Lowering Therapy

The results of a dozen such studies were reported between 1984 and 1994, most of them from the USA.[2] These ranged in size from less than 50 to over 800 subjects and lasted between 1 and 10 years. Most involved a randomized comparison of diet versus diet combined with lipid-lowering drugs. In the earlier trials, bile acid sequestrants, such as cholestyramine and colestipol, were given alone or in combination with either nicotinic acid or lovastatin. In later trials, lovastatin or simvastatin were given as monotherapy. Alternative forms of treatment included partial ileal bypass, exercise and anti-stress measures. In the earlier trials, angiographic changes were assessed visually, whereas the later ones used computer-assisted quantitative coronary angiography (QCA).

Baseline levels of serum total and LDL cholesterol varied considerably between the 12 trials, with overall means of 6.3 and 4.4 mmol/l, respectively. During the trials, LDL cholesterol was 31% lower and HDL cholesterol, 5% higher in the intervention groups than in the controls. Angiographic criteria varied, but all patients were categorized according to whether their coronary lesions showed progression, regression, mixed response or no change. Of the more than 2,500 patients involved, 44% of the controls were classified as progressors and 9% as regressors, compared with 29% and 18%, respectively, of those on treatment. Hence, lipid-lowering treatment reduced the chances of progression by one-third and increased the chances of regression twofold.

Although not powered to detect changes in clinical end points, three trials[3-5] did show significant reductions both in cardiovascular events and in the need for revascularisation procedures. Greg Brown and colleagues in Seattle examined the relation between clinical events and coronary artery plaque size in one of these trials, the Familial Atherosclerosis Treatment Study (FATS),[4] and concluded that mild to moderately stenosed lesions were usually the culprits and were more responsive to lipid-lowering therapy than were severe (>70% stenosed) lesions.[6]

As shown by the late Michael Davies and his colleagues at St. George's Hospital in London, high-grade stenotic lesions contain large amounts of collagen and calcium but relatively little lipid, whereas the propensity of plaques to undergo fissuring and thrombosis, and thereby precipitate a clinical event, is determined by the amount of extracellular lipid in the core of the plaque.[7] Thus, a reduction in the amount of lipid in moderately stenotic lesions, resulting in only a modest decrease in plaque size, would be expected to be accompanied by an increase in plaque stability and a reduction in clinical events, as occurred in FATS.

Although the relatively slight quantitative changes in vessel lumen in the regression trials were probably haemodynamically insignificant, they seem to have been an important marker of plaque stabilization. This being the case, the next step was to investigate the effect of statins on clinical outcomes in trials of sufficient size to achieve statistically significant results. The first of these was the Scandinavian Simvastatin Survival Study (4S).

17.3. Pedersen: Biography

The Norwegian physician Terje Pedersen was the initiator, organiser and lynchpin of the 4S (Fig. 17.2). He was born in 1945, graduated from the

Fig. 17.2.   Terje Pedersen (Reprinted from Morris K. Terje Pedersen: A pioneer trialist in preventive cardiology. *The Lancet* 2010; 375: 717, with permission from Elsevier).

local high school in 1965 and qualified in Medicine at Heidelberg University, Germany in 1972. He did his internship in a small Norwegian hospital situated above the Arctic Circle, obtained a PhD from the University of Bergen based on his participation in a clinical trial of a β-blocker and qualified as a cardiologist in 1986. The following year, during a conversation with the Director of Merck Scandinavia, he learned that the company had a novel cholesterol-lowering compound, simvastatin. He persuaded the Director of the desirability of conducting a trial of this drug with total mortality as the clinical outcome and they put forward their proposal to Merck in Rahway, to which Merck eventually agreed.

Pedersen, the Director of Merck Norway and the Director of Merck Scandinavia between them decided upon a multi-centre trial involving 94 sites in Norway, Sweden, Denmark, Finland and Iceland with Pedersen as Scientific Coordinator. This involved recruiting 7,000 patients with a serum cholesterol of 5.5–8.0 mmol/l and evidence of coronary heart disease, of whom 4,444 were eventually randomised to simvastatin or placebo.

During the five and a half years that the trial lasted, regular scientific meetings were held for participating physicians and the author was invited to speak at one of these events held in Bergen. After presenting some coronary angiographic data showing regression in FH patients treated with apheresis and lovastatin he suggested that this provided anecdotal evidence that LDL-lowering was beneficial. Pedersen came down on him like a ton of bricks in the discussion and said that the only real proof of benefit from cholesterol-lowering was in the prevention of fatal events — he was determined to ensure that all his investigators and their patients stuck to the protocol right up until the end of the trial, still a few months hence!

In the autumn of 1994, Pedersen presented the results of 4S at the annual meeting of the American Heart Association, and they were published simultaneously in *The Lancet*.[8] The trial generated enormous professional and public interest and Pedersen estimated that he gave presentations in 72 cities worldwide during 1995. In 1998, he was appointed Professor of Clinical Cardiology at the University of Oslo. His hobbies include cross-country skiing and fly-fishing.[9]

17.4. The Scandinavian Simvastatin Survival Study (4S)

The results of 4S provided unequivocal evidence that treatment of hypercholesterolaemia in patients with existing coronary heart disease reduced both cardiovascular events and total mortality.[8] In this epic secondary prevention trial, 4,444 mostly male patients from five Scandinavian countries, aged 35–70 with angina or previous myocardial infarction and with serum cholesterol averaging 6.8 mmol/l on diet, were randomized to receive simvastatin or placebo. Simvastatin dosage was initially 20 mg/day but was increased to 40 mg/day as necessary, so as to maintain serum cholesterol in the range 3–5.2 mmol/l. During the study, simvastatin reduced total and LDL cholesterol by 25% and 35%, respectively, and increased HDL cholesterol by 8%.

The median duration of treatment was 5.4 years, after which the trial was stopped because of a highly significant 30% reduction in total mortality. This was entirely due to a 42% reduction in coronary mortality, and there was no increase in non-cardiovascular causes of death, including cancer and trauma. In addition to its effects on coronary disease, simvastatin significantly reduced the occurrence of cerebrovascular events. The mean level of LDL cholesterol in subjects on simvastatin in 4S was 3.2 mmol/l, similar to the 3 mmol/l seen in the earlier angiographic trials of lipid-lowering therapy.[2]

Taken together, the findings of the angiographic trials together with those from 4S clearly demonstrated that a 30–35% reduction in LDL cholesterol resulted in decreased progression of coronary atheromatous lesions and fewer clinical events; the 4S authors estimated that simvastatin had prevented over 40% of the deaths that otherwise would have occurred. 4S cost Merck about $45 million, but on the credit side, annual sales of simvastatin were projected to be in the region of $4,000 million in 1998. Both in scientific and commercial terms 4S was an outstanding success, and much of the credit for this is due to Terje Pedersen.

17.5. Shepherd: Biography

Jim Shepherd, the principal investigator of the West of Scotland Coronary Primary Prevention Study (WOSCOPS), was born in 1944 in Scotland and educated at the Hamilton Academy in South Lanarkshire. From there he went to the University of Glasgow where he obtained a BSc in 1965, qualified MB BCh with honours in 1968 and got a PhD in 1972.

He was lecturer in Biochemistry between 1969 and 1972, then senior lecturer until 1988 when he was appointed Professor and head of the Department of Vascular Biochemistry. In 2006, he became Emeritus Professor of Cardiovascular and Medical Sciences of the University of Glasgow. He was a Fellow of the Royal Society of Edinburgh and a Fellow of the Academy of Medical Sciences. His hobbies included fishing and Scottish dancing, both of which he actively pursued in Scotland.

In 1985, he convened a meeting of a "Lipid Clinics Group" at the Hammersmith Hospital, where the author was the local organiser. The meeting was attended by a score of physicians and clinical biochemists in charge of lipid clinics in the UK and resulted in the formation of the British Association of Lipid Disorders (BALD), with Jim as chairman. This was renamed the British Hyperlipidaemia Association (BHA) the following year and became HEART UK in 2003 (Fig. 17.3). He was also President of the European Atherosclerosis Society from 1993–1996.

17.6. The West of Scotland Coronary Primary Prevention Study (WOSCOPS)

According to Chris Packard,[10] WOSCOPS[11] was conceived by Jim Shepherd in 1988 during a visit to the Glasgow Royal Infirmary by a

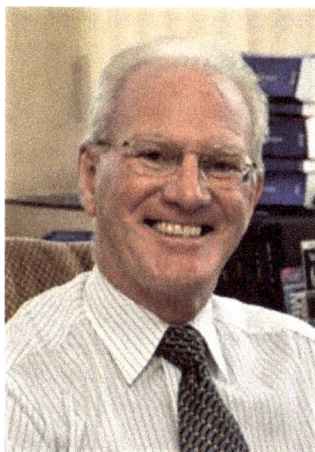

Fig. 17.3. Jim Shepherd (1944–2022). (Photo: By courtesy of the University of Glasgow).

senior director of Squibb, the licensee in Britain of pravastatin. At that meeting, the local cardiologists voted for a secondary prevention trial with pravastatin, but Jim Shepherd persuaded Squibb that there was a greater need for a primary prevention trial. Squibb decided to back his rather than the cardiologists' proposal, with the result that WOSCOPS became the first primary prevention trial to use a statin. This initially involved screening more than 80,000 men in the West of Scotland, of whom 6,595 were selected on the basis of being asymptomatic, aged 45–64, with a serum total cholesterol of over 6.5 mmol/l and an LDL cholesterol of 4.5–6 mmol/l. They were then randomised to receive pravastatin 40 mg/day or a placebo for an average duration of just under 5 years.

Pravastatin decreased serum total and LDL cholesterol by 20% and 26%, respectively, and raised HDL cholesterol by 5%. These changes were associated with a highly significant 31% reduction in non-fatal myocardial infarcts plus coronary deaths. Although the 22% reduction in total mortality just failed to achieve statistical significance, there was no increase in non-cardiovascular causes of death in pravastatin-treated subjects. The results were presented to the American Heart Association and published simultaneously in the *New England Journal of Medicine* in 1995, exactly a year after the publication of 4S. Like the latter, WOSCOPS marked an important milestone in the management of hyperlipidaemia and prevention of cardiovascular disease and much of its success was due to Jim Shepherd's innovative approach to research, his organisational skills and his effective interpersonal relationships.[12] He died in 2022, aged 78.

17.7. Collins: Biography

Rory Collins was born in 1955, educated at Dulwich College and studied medicine at St. Thomas's Hospital Medical School, London, qualifying there in 1980. He also studied statistics at George Washington University while he was an undergraduate and subsequently at Oxford University as a postgraduate. In 1985, he was appointed co-director of the University of Oxford's Clinical Trials Unit and in 1996 he became Professor of Medicine and Epidemiology (Fig. 17.4). He is also the principal investigator and chief executive of the UK Biobank study. He is a Fellow of the Royal Society (FRS) and of the Academy of Medical Sciences. He was the principal investigator of the Heart Protection Study (HPS) and helped design and was co-author of the Cholesterol Treatment Trialists' meta-analysis of the efficacy of statins in reducing adverse cardiovascular events. Like Jonathan Tobert he has also investigated and helped validate the safety of these compounds. He was knighted in 2011 for services to science.

Fig. 17.4.   Sir Rory Collins (Photo: By courtesy of John Cairns/Oxford Population Health).

17.8. The Heart Protection Study (HPS)

In addition to 4S and WOSCOPS several other statin trials were published between 1995 and 1998, but the largest of them was not completed until the start of the new millennium. This was the MRC/BHF Heart Protection Study (HPS),[13] which was designed and conducted by the Oxford-based Clinical Trials Service Unit, with Rory Collins as the principal investigator. This huge trial examined the effects on mortality and morbidity of cholesterol-lowering therapy in more than 20,000 subjects with, or at high risk of, cardiovascular disease in the UK. Men and women aged 40–80 with serum total cholesterol over 3.5 mmol/l were randomised to receive either simvastatin 40 mg daily or anti-oxidant vitamins, the two combined, or placebo.

The results showed an incidence of major coronary events in those on placebo of 11.8% over 5 years, confirming that they were a high-risk group. Subjects allocated to simvastatin had a mean reduction in LDL cholesterol of 1 mmol/l, with decreases in total and cardiovascular mortality of 12% and 17% and decreases in coronary events and strokes of 26% and 27%, respectively. Benefit from simvastatin occurred irrespective of the level of LDL cholesterol at entry to the study, and was not influenced by age, gender or clinical status, and there was no increase in non-cardiovascular mortality. One-third of the patients in the HPS had a baseline LDL cholesterol under 3 mmol/l, an important finding suggesting that high-risk individuals benefit from LDL-lowering even if their LDL cholesterol is regarded as being "normal."

As reported in a concomitant publication in *The Lancet*,[14] the HPS not only assessed the efficacy of simvastatin in reducing cardiovascular events, but it also investigated the efficacy in that respect of an anti-oxidant cocktail of vitamin E 600 mg, vitamin C 250 mg and β-carotene 20 mg administered daily over a period of 5 years. Despite doubling the plasma concentration of α-tocopherol and significantly increasing levels of vitamin C and β-carotene, there were no differences in total mortality, myocardial infarction or stroke between subjects given anti-oxidants compared with those given a placebo. The lack of any benefit from anti-oxidants was equally evident in diabetics, who are known to be under increased oxidative stress. These results from HPS were included in the meta-analysis referred to in Chapter 12, which concluded that anti-oxidant supplementation should not feature in cardiovascular disease risk-reduction strategies.

17.9. The Cholesterol Treatment Trialists' Collaboration

The final dotting of *i*'s and crossing of *t*'s that marked the end of the controversy over the lipid hypothesis came when data from over 90,000 individuals who participated in 14 statin trials in 1994–2004 were meta-analysed by Colin Baigent and his colleagues in the Cholesterol Treatment Trialists' Collaboration.[15] The results showed decreases of 12% and 19% in total and coronary mortality, respectively, for each 1 mmol/l reduction in LDL cholesterol, similar to what was found in the HPS. Overall, the risk of a major vascular event, including strokes, was reduced by one-fifth. No significant changes in non-cardiovascular mortality occurred over the five-year duration of these trials, and it is probable that even McMichael and Mitchell would have been impressed by these data had they still been alive. Another more recent analysis by the Cholesterol Treatment Trialists' Collaboration showed that <10% of muscle symptoms reported by participants in 19 double-blind statin trials were due to statins, the risk being much lower than the known cardiovascular benefits.[16]

Which brings us back to 1953, when Oliver and Boyd first presented their findings showing an association between raised cholesterol and coronary disease in British men. That same year the American physician O.J. Pollak published a paper describing the cholesterol-lowering properties of plant sterols, which concluded with the prophesy "Someday, the question as to the value of prophylactic or therapeutic reduction of blood cholesterol will be answered."[17] He was right, more than 50 years later it has been — in the affirmative on both counts.

References

1. Blankenhorn DH, Azen SP, Kramsch DM, *et al.* Coronary angiographic changes with lovastatin therapy. The Monitored Atherosclerosis Regression Study (MARS). *Ann. Intern. Med.* 1993; **119**: 969–976.
2. Thompson GR. Angiographic trials of lipid-lowering therapy: End of an era? *Br. Heart J.* 1995; **74**: 343–347.
3. Buchwald H, Varco RL, Matts JP, *et al.* Effect of partial ileal bypass surgery on mortality and morbidity from coronary heart disease in patients with hypercholesterolemia. *N. Engl. J. Med.* 1990; **323**: 946–955.
4. Brown G, Albers JJ, Fisher LD. Regression of coronary artery disease as a result of intensive lipid-lowering therapy in men with high levels of apolipoprotein B. *N. Engl. J. Med.* 1990; **323**: 1289–1298.

5. Watts GF, Lewis B, Brunt JNH, *et al.* Effects on coronary artery disease of lipid-lowering diet, or diet plus cholestyramine in the St. Thomas' Atherosclerosis Regression Study (STARS). *Lancet* 1992; **339**: 563–569.

6. Brown BG, Zhao X-Q, Sacco DE, Albers JJ. Atherosclerosis regression, plaque disruption, and cardiovascular events. A rationale for lipid lowering in coronary artery disease. *Ann. Rev. Med.* 1993; **44**: 365–376.

7. Davies MJ, Krikler DM, Katz D. Atherosclerosis: Inhibition or regression as therapeutic possibilities. *Br. Heart J.* 1991; **65**: 302–310.

8. Scandinavian Simvastatin Survival Study Group. Randomised trial of cholesterol lowering in 4444 patients with coronary heart disease: The Scandinavian Simvastatin Survival Study (4S). *Lancet* 1994; **344**:1383–1389.

9. Pedersen, TR. Terje Rolf Pedersen, MD: A conversation with the Editor. *Am. J. Cardiol.* 1999; **84**: 1234–1245.

10. Packard C. Cholesterol, atherosclerosis and coronary disease in the UK, 1950–2000. Vol. 27: p. 77, Reynolds LA, Tansey EM (eds.). London: Wellcome Trust Centre for the History of Medicine; 2006.

11. Shepherd J, Cobbe SM, Ford I, *et al.* Prevention of coronary heart disease with pravastatin in men with hypercholesterolemia. West of Scotland Coronary Prevention Study Group. *N. Engl. J. Med.* 1995; **333**: 1301–1307.

12. Packard CJ. In Memoriam: Professor James Shepherd FRSE. Atherosclerosis 2022. doi: 10.1016/j.atherosclerosis.2022.06.1021.

13. Heart Protection Study Collaborative Group. MRC/BHF heart protection study of cholesterol lowering with simvastatin in 20536 high-risk individuals: A randomised placebo-controlled trial. *Lancet* 2002; **360**: 7–22.

14. Heart Protection Study Collaborative Group. MRC/BHF Heart Protection Study of antioxidant vitamin supplementation in 20536 high-risk individuals: A randomized placebo-controlled trial. *Lancet* 2002; **360**: 23–33.

15. Baigent C, Keech A, Kearney PM, *et al.* Efficacy and safety of cholesterol-lowering treatment: Prospective meta-analysis of data from 90,056 participants in 14 randomised trials of statins. *Lancet* 2005; **366**: 1267–1278.

16. Cholesterol Treatment Trialists' Collaboration. Effect of statin therapy on muscle symptoms: An individual participant data meta-analysis of large-scale, randomized, double-blind trials. *Lancet.* Published online August 29, 2022. doi: 10.1016/S0140-6736(22)01545-8.

17. Pollak OJ. Reduction of blood cholesterol in man. *Circulation* 1953; **7**: 702–706.

Chapter 18

Impact of the Lipid Hypothesis on the Science and Practice of Medicine

18.1. Summary of the Evidence that Proved the Lipid Hypothesis

A recap of the preceding chapters reveals an almost continuous chain of discoveries, punctuated by occasional setbacks, upon which the lipid hypothesis rests. Anitschkow initiated this sequence of events when he induced atheroma in rabbits rendered hypercholesterolaemic by feeding them large amounts of cholesterol in their diet. Approximately 40 years and two world wars later Gofman, and subsequently Havel, showed that cholesterol was transported in the blood as a constituent of lipoprotein particles and that hypercholesterolaemic rabbits, patients with familial hypercholesterolaemia (FH) and survivors of myocardial infarction all had raised levels of LDL.

Dawber and Kannel and their long-running Framingham Study confirmed the role of a raised level of LDL cholesterol as a risk factor for atherosclerotic cardiovascular disease (ACVD) in a population setting. They also confirmed the finding of Miller and Miller that a low level of HDL cholesterol was another risk factor for ACVD. Concurrently, Ancel Keys's controversial Seven Countries Study showed that differences in serum cholesterol levels between countries such as Finland and Japan reflected differences in the saturated fat content of their diets, a finding that continues to influence most dietary guidelines to this day.

Michael Oliver then entered the fray and raised doubts over the safety of lowering serum cholesterol, based on the WHO trial of clofibrate, which showed that this lipid-lowering compound prevented non-fatal myocardial infarcts but increased total mortality. The hitherto rather negative British influence became evident again in the shape of the nutritionist John Yudkin, who believed that sugar not saturated fat was atherogenic, and the cardiologists McMichael and Mitchell who harboured deep-rooted but unsubstantiated objections to the lipid hypothesis. Myant helped restore progress with his clinical research studies of patients with homozygous FH and support for the introduction of apheresis to investigate and treat patients with this disorder.

The publication by Fredrickson, Levy and Lees of their classification of lipoprotein phenotypes in 1967 heralded the birth of clinical lipidology. This set the scene for the classical studies of Goldstein and Brown that earned them the Nobel Prize in 1985. They discovered not only the LDL receptor and showed that inherited defects of the latter caused FH, but they also discovered the scavenger receptor. Steinberg then showed that oxidised LDL, but not native LDL, was taken up avidly by this scavenger receptor and he suggested that macrophage-mediated oxidation of LDL and its subsequent uptake via the scavenger receptor pathway played a key role in atherogenesis. However, the corollary that treatment with anti-oxidants prevents atherosclerosis in humans remains unproven.

Berg's discovery of Lp(a) revealed an enigmatic lipoprotein without any defined physiological function but which is a heritable and hard-to-treat risk factor for cardiovascular disease when it is present in excess. In contrast, Utermann's discovery of apoE revealed a functionally important apoprotein that binds to the apoB/E or LDL receptor, thereby enabling uptake of triglyceride-rich remnant particles by the liver. It exists as three major isoforms, the commonest one being apoE3, whereas inheritance of apoE2 can cause type III hyperlipoproteinaemia and inheritance of apoE4 is associated with an increased risk of developing Alzheimer's disease.

The penultimate three chapters are all devoted to the statin era, starting with Endo's momentous discovery of compactin. This was followed by the discovery and initially somewhat fraught development of lovastatin and subsequently of simvastatin for clinical use as cholesterol-lowering agents by Vagelos, Alberts and Tobert at Merck. Pedersen, Shepherd and Collins then demonstrated the ability of simvastatin and pravastatin to reduce the morbidity and mortality of subjects with or at increased risk of ACVD by lowering their LDL cholesterol. Finally, the meta-analysis by the Cholesterol Treatment Trialists' Collaboration of the 90,000 individuals who participated in statin trials confirmed the validity of the lipid hypothesis beyond all reasonable doubt.

18.2. Consequences of Validating the Lipid Hypothesis

Four consequences of validating the Lipid Hypothesis have been: (1) advances in research into the pathogenic mechanisms whereby cholesterol causes atherosclerosis; (2) a huge boost to the production of clinical guidelines on the management of dyslipidaemia; (3) the widespread establishment of hospital outpatient lipid clinics; (4) stimulation of the development and testing of novel lipid-lowering compounds.

18.2.1. *Current concepts of atherogenesis*

Peter Libby has been foremost in proposing the concept that raised or even "normal" levels of plasma LDL initiate a series of inflammatory reactions that are responsible for atherogenesis, the evidence for which he summarised in 2002.[1] This sequence of events starts with the attachment of mononuclear leucocytes to endothelial cells in the arterial intima, promoted by the endothelium-derived vascular cell adhesion molecule (VCAM-1). The latter is induced by the accumulation of oxidised LDL in the arterial intima and results in the production of pro-inflammatory cytokines (proteins involved in cell signalling) such as interleukin-1β (IL-1β) and tumour necrosis factor α (TNF-α). Recruitment of monocytes into the arterial intima is enhanced by these cytokines and by monocyte chemoattractant protein -1 (MCP-1).

Once in the arterial intima, monocytes acquire the morphological characteristics of macrophages, promoted by monocyte colony-stimulating factor (M-CSF). These macrophages then express scavenger receptors for oxidised or otherwise modified LDL, become enriched in cholesterol esters and turn into foam cells. The latter are the characteristic feature of early atherosclerosis, first noted by Anitschkow. In situations where extreme hypercholesterolaemia is the atherogenic stimulus, such as in homozygous FH, macrophages accumulate such large quantities of cholesterol ester that they undergo necrosis and release enzymes that hydrolyse cholesterol esters, resulting in the formation of extracellular deposits of cholesterol crystals in the base of the plaque. HDL appears to protect against plaque formation, probably via its role in reverse cholesterol transport.

In addition to foam cells, the other main constituent of atheromatous plaques are smooth muscle cells, which synthesise collagen in response to platelet-derived transforming growth factor-β (TGF-β) and platelet-derived growth factor (PDG-F) and transform fatty streaks into fibro-fatty plaques, with a lipid core and a fibrous cap. The likelihood that such plaques might rupture and precipitate an ischaemic event is greatest if they have an extracellular lipid-rich core covered by a thin fibrous cap.

In 2021, Libby updated these concepts by firstly, downgrading the protective role of HDL in atherosclerosis and instead emphasising the atherogenic role of triglyceride-rich lipoproteins and their remnants, with which HDL levels are inversely correlated; second, by proposing that superficial erosions of the vessel wall are a commoner cause of arterial thrombosis and hence of clinical events than rupture of vulnerable plaques; and third, by substituting native or aggregated LDL for oxidised LDL as the initiator of atherosclerosis.[2]

Much of our current knowledge of atherosclerosis is based therefore upon the validity of the lipid hypothesis and the causal role of LDL cholesterol in initiating fatty streaks and the subsequent inflammatory process that leads to plaque formation, especially in hypercholesterolaemic subjects. Statins prevent or reverse this process by lowering LDL cholesterol but also perhaps by exerting anti-inflammatory effects that are independent of LDL-lowering. Non-statin drugs with an anti-inflammatory action such as colchicine may have a secondary role to play in treating atherosclerotic vascular disease unresponsive to LDL-lowering therapy.

18.2.2. *Guidelines on lipids in the prevention of cardiovascular disease*

A search of PubMed at the end of 2021 using the term "guidelines, lipids and cardiovascular disease" revealed 4,276 publications since 1979. There were four publications in 1979, 27 in 1994 (the year 4S was published), 99 in 2000 and then a steady increase each year up to its zenith in 2014 when there were 298 publications. Thereafter they averaged 247 publications per year until 2021, as shown in Fig. 18.1.

Not all these publications describe guidelines, but since the 1980s there have been significant changes in the recommendations made by

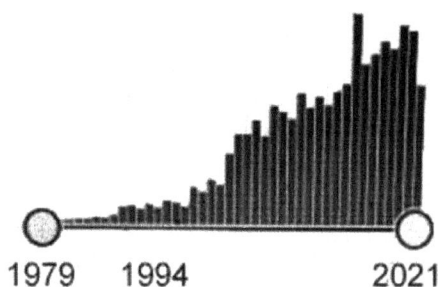

Fig. 18.1. Relative frequency of publications on "guidelines, lipids and cardiovascular disease" between 1979–2021. The peak year was 2014 with almost 300 publications listed in PubMed.

various organisations in numerous countries. In 1984, the NHLBI convened a Consensus Development Conference on Lowering Blood Cholesterol to Prevent Heart Disease in the USA under the chairmanship of Dan Steinberg.[3] Its recommendations included defining levels of serum cholesterol considered to be associated with a moderate or high risk of coronary heart disease, which in ≥40-year-olds were >240 mg/dl (6.2 mmol/l) and >260 mg/dl (6.7 mmol/l), respectively.

Treatment of hypercholesterolaemia initially consisted of a cholesterol-lowering diet, with recourse to drug therapy only if diet failed to lower the cholesterol to <200 mg/dl (5.2 mmol/l). In those days the only drugs available were anion exchange resins, fibrates and nicotinic acid. Similar recommendations were subsequently proposed by the Adult Treatment Panel of the National Cholesterol Education Program (NCEP) in the USA,[4] by the British Hyperlipidaemia Association,[5] by the European Atherosclerosis Society[6] and by analogous organisations in Canada,[7] New Zealand[8] and South Africa.[9]

The NCEP guidelines classified serum total cholesterol levels as desirable (<200 mg/dl, <5.2 mmol/l), borderline-high (200–240 mg/dl, 5.2–6.2 mmol/l) and high (≥240 mg/dl, ≥6.2 mmol/l), with diet as the primary cholesterol-lowering therapy. They identified high levels of LDL-cholesterol as the primary target for cholesterol-lowering therapy and low levels of HDL-cholesterol as an additional risk factor for coronary heart disease. Failure of diet after 6 months was regarded as an indication of the need for drug therapy. Similar criteria were adopted by the British and European guidelines, both of which advocated a target level of total cholesterol of <200 mg/dl (<5.2 mmol/l) on treatment.

More than 30 years and numerous guidelines later, the modalities and targets of treatment in the statin era had altered radically. Thus, the 2018 American College of Cardiology/ American Heart Association guidelines[10] recommend that adults aged 40–75 with an LDL cholesterol of ≥70 mg/dl (≥1.8 mmol/l) and a 10-year risk of ASCVD of ≥7.5% are eligible for treatment with a moderate intensity statin. Those with ASCVD or at high risk of same, including patients with primary hypercholesterolaemia (LDL cholesterol ≥190 mg/dl, ≥4.9 mmol/l) or diabetes, should be treated with a moderate or high intensity statin, supplemented with ezetimibe if the LDL cholesterol remains ≥100 mg/dl (≥2.6 mmol/l) or ≥70 mg/dl (≥1.8 mmol/l) in very high risk subjects. Risk-enhancing factors include an Lp(a) of ≥50 mg/dl or 115 nmol/l. The objective of treatment is to lower the LDL cholesterol by ≥50%, but in the event that this is not achieved the addition of a PCSK9 inhibitor is recommended.

The main message of these contemporary guidelines is that LDL cholesterol plays a fundamental role as a risk factor for ASCVD and that the more it is lowered, the greater is the reduction in risk. A similar message was conveyed by the updated European Society for Cardiology/European Atherosclerosis Society guidelines in 2019[11] and by the even more radical consensus statement published in 2020 by the Lipid Association of India.[12]

18.2.3. *The establishment of lipid clinics*

Probably the most famous if not the first lipid clinic in the world was established at the NHLBI in Bethesda by Fredrickson and colleagues in the early 1960s. The first outpatient lipid clinic in the UK was set up at the Hammersmith Hospital in the mid-1960s and 10 years later it became the first centre in the country to which patients with severe hypercholesterolaemia could be referred for apheresis. Currently, in 2022, there are 28 lipid clinics within a radius of 10 miles from the centre of London and numerous others throughout the UK, eight of which undertake lipoprotein apheresis.

Facilities vary, but the majority of clinics in the UK provide diagnostic and therapeutic advice, including access to genetic testing for inherited disorders of lipoprotein metabolism such as FH and type III hyperlipidaemia. This entails sending samples to specialised laboratories equipped to sequence DNA for pathogenic mutations in the genes encoding the LDL receptor, apoB, PCSK9 and apoE, although some hospital laboratories do perform apoE genotyping. Lipid clinics also play an important role in organising cascade screening of relatives of FH patients and in providing pharmaceutical companies with access to patients wishing to participate in clinical trials of novel lipid-lowering compounds.

The majority of lipid clinics no longer undertake the laboratory investigations such as agarose gel electrophoresis and preparative ultracentrifugation of lipoproteins needed to establish a phenotypic diagnosis in dyslipidaemic patients. However, a new approach to lipoprotein phenotyping based solely on routinely measured values of non-HDL cholesterol and triglycerides has recently been proposed.[13] In addition to performing a full lipid profile, most lipid clinics now regard levels of apoA-1, apoB and Lp(a) as important indices of CVD risk and measure them routinely in newly referred patients.

18.2.4. *Statins, ezetimibe and novel forms of lipid-lowering therapy*

This penultimate chapter reappraises the current role of statins in the prevention and treatment of atherosclerotic cardiovascular disease and describes the remarkable advances in lipid-lowering therapy during the 21st century, which are predicated upon the need to maximise reduction of LDL. The therapeutic revolution that took place is analogous to the one that followed the introduction of antibiotics after the Second World War and was initiated in the 1970s by the discovery of the LDL receptor and promulgated by the discovery of compactin and its successors. Statins remain first-line treatment for most patients with or at high risk of cardiovascular disease, but the post-statin era of novel lipid-lowering agents holds great promise for the management of severe lipid disorders that are refractory to statins, especially homozygous FH.

18.2.4.1. *Statins and ezetimibe*

A survey conducted in 1999 by Andrew Neil and the Simon Broome Register Group, involving more than 1,000 UK patients with the less severe, heterozygous form of FH showed that their mortality from coronary heart disease had halved since 1992.[14] This was attributed to the widespread use of statins to treat this particular group of patients from

1989 onwards, when simvastatin was first licensed in Britain. One possible explanation for the magnitude of this decrease in mortality over such a short space of time is that FH heterozygotes, whose main risk factor is a raised LDL, respond better to lipid-lowering therapy than do non-FH patients with a lower LDL cholesterol but with other contributory risk factors.[15]

The majority of patients with hyperlipidaemia, including those with FH, respond adequately to statin therapy alone or combined with the cholesterol absorption blocker ezetimibe. Addition of the latter results in an additional 15% reduction in LDL cholesterol and enables a lower dose of statin to be used if desired. However, there is a minority of subjects whose LDL cholesterol remains high despite intensive statin therapy or who claim to be intolerant of these drugs (see Chapter 15).

Recently, it has become apparent that statins may induce diabetes in some individuals, especially if they already have impaired glucose tolerance. The mechanism of this effect is uncertain but seems to be related to the potency and dose of the statin concerned, perhaps reflecting the extent to which the latter decreases cholesterol synthesis in insulin-secreting pancreatic islet cells. Hence, in subjects with or at risk of developing diabetes, it is prudent to use as low a dose as feasible of a low intensity statin, supplemented if necessary with ezetimibe, which is not diabetogenic.

18.2.4.2. *PCSK9 inhibitors*

For genuinely statin-intolerant individuals and for patients with statin-refractory FH, a new class of lipid-lowering agents has recently become available. These are the PCSK9 inhibitors, monoclonal antibodies that reduce the rate of degradation of LDL receptors by PCSK9 and thereby promote receptor-mediated LDL uptake by the liver; these are given by injection every 2 to 4 weeks. Their LDL-lowering effect is equivalent to that of a high dose, potent statin and is additive to the effect of the latter.

The discovery of PCSK9 inhibitors is a fascinating story and, like the discovery of the LDL receptor, resulted from studies of patients with rare genetic mutations. Proprotein convertase subtilisin kexin 9, to give PCSK9 its full name, is the ninth member of a family of human proteolytic enzymes, one of its functions being to induce the lysosomal degradation (biological destruction) of the LDL receptor. In 2003, hypercholesterolaemia due to an increase in LDL levels was reported in members of two French families with a gain of function mutation of PCSK9.[16] Two years later it was reported that loss-of-function mutations of the gene were associated with subnormal levels of LDL and a markedly reduced risk of coronary heart disease.[16] This pointed to the potential of PCSK9 inhibition as a means of treating hypercholesterolaemia, either as a substitute for statins in statin-intolerant subjects or as an adjunct in statin-refractory subjects, given that the ability of statins to upregulate the LDL receptor and thereby lower LDL is limited by the increase in PCSK9 levels that they induce.[17]

Current pharmacological approaches to inhibiting PCSK9 are humanised monoclonal antibodies (mAbs), two of which, evolocumab and alirocumab, are licensed for use in patients with or at high risk of developing atherosclerotic CVD. A recent meta-analysis of 24 randomised trials in over 10,000 subjects given various doses of evolocumab or alirocumab by subcutaneous injection once or twice monthly showed a mean reduction in LDL cholesterol of 47%.[18] Data from individual trials of evolocumab showed reductions in LDL cholesterol of 60% at maximum dosage; mAb-related decreases were similar irrespective of whether patients did or did not have heterozygous FH[19,20] or of whether or not they were receiving concomitant statin therapy.[21] Two open label trials of evolocumab showed that doses of 140 mg every 2 weeks or 420 mg monthly for 11 months reduced LDL cholesterol by 61% from 3.1 to 1.2 mmol/l and halved the risk of CV events.[22]

In the meta-analysis referred to above, levels of Lp(a) were reduced by 25% which is roughly half the reduction achieved in LDL cholesterol. This illustrates one of the limitations of PCSK9 mAbs and probably reflects the fact that Lp(a) is not catabolised by the LDL receptor pathway. Another limitation is their relatively low efficacy in patients with homozygous FH, reductions in LDL cholesterol on evolocumab averaging only 23%.[23]

From the practical point of view, PCSK9 mAbs clearly have enormous potential as an adjunct to statins and ezetimibe in the treatment of patients with heterozygous FH and as an alternative in statin-refractory or intolerant patients with CVD. However, their usefulness in homozygous FH is much less and often they will complement rather than replace lipoprotein apheresis.

The most recent development in the field of PCSK9 inhibition is Inclisiran, a small interfering RNA (siRNA) nucleotide molecule that prevents the translation of PCSK9 in the liver.[24] Given twice yearly it achieves a 50% reduction in LDL cholesterol over and above the effect of background statin/ezetimibe therapy and has recently been approved by the FDA for use in the treatment of heterozygous FH. The availability of the twin LDL receptor-upregulating options of statins and PCSK9 inhibitors means that raised LDL cholesterol levels can be lowered effectively in the majority of patients exhibiting some degree of residual LDL receptor activity.

18.2.4.3. *Lomitapide and Evinacumab*

Lomitapide, a microsomal triglyceride transfer protein (MTP) inhibitor, is used to treat extreme forms of hypercholesterolaemia such as homozygous FH,[25] and hypertriglyceridaemia, specifically familial hyperchylomicronaemia. Unlike all the other lipid-lowering drugs described above, lomitapide inhibits the secretion of apoB-containing lipoproteins and does not depend upon LDL receptor activity. In essence it induces an iatrogenic form of aβlipoproteinaemia (recessively inherited MTP deficiency) and has the latter's drawback of causing a fatty liver, but it remains to be seen whether this will eventually lead to hepatic fibrosis.

The latest therapeutic advance is Evinacumab, a monoclonal antibody to angiopoietin-like 3 (ANGPTL3) that has been shown in a preliminary study to decrease LDL cholesterol by 49% in FH homozygotes.[26] Other studies show that the mechanism is independent of the LDL receptor pathway and involves increased activity of lipoprotein lipase. Evinacumab does not induce a fatty liver, but further studies are necessary before its therapeutic potential and safety can be properly evaluated.

18.3. Has the Controversy Over the Lipid Hypothesis Been Finally Resolved?

Forty years ago, Nick Myant, a scrupulously honest scientist, commented on the need for objectivity when considering the role of cholesterol in ischaemic heart disease: "It should not be necessary to refer here to the need for balance and objectivity in the sifting of evidence, something that has been taken for granted by the scientific community for about 300 years. Nevertheless, it is a fact that the very word "cholesterol" tends to arouse emotional reactions that cloud judgement … The reason for this state of affairs is, of course, the link thought to exist between cholesterol and a disease so serious and widespread that it is difficult not to feel strongly about it."[27]

The controversy over the Lipid Hypothesis may have died down, but it is not extinct. As Francis Bacon pointed out over 400 years ago, "If a man will begin with certainties, he shall end in doubts; but if he will be content to begin with doubts, he shall end in certainties."[28] However, there will always be some who begin with doubts and never relinquish them, epitomised by the International Network of Cholesterol Sceptics. They put into practice their motto "The growth of knowledge depends entirely on disagreement," by publishing letters and review articles with pejorative titles such as "LDL-C does not cause cardiovascular disease."[29] In these communications they criticise others for falsifying the lipid hypothesis by using misleading statistics and ignoring contradictory findings, but they never provide any experimental data supporting an alternative explanation for the aetiology of atherosclerosis.

Currently, in the middle of the COVID pandemic, there is a striking similarity in the attitudes of the anti-vaccination lobby and of the Cholesterol Sceptics, but in both instances their views have been comprehensively rebutted. To conclude by quoting Nick Myant again, "I see no advantage in adopting an evangelistic approach, since the question will ultimately be settled by the methods of science."[27] He was right — it has been.

References

1. Libby P. Inflammation in atherosclerosis. *Nature* 2002; **420**: 868–874.
2. Libby P. The changing nature of atherosclerosis: What we thought we knew, what we think we know, and what we have to learn. *Eur. Heart J.* 2021; **42**: 4781–4790.
3. Lowering blood cholesterol to prevent heart disease. NIH consensus development conference statement. *Arteriosclerosis* 1985; **5**: 404–412.
4. National cholesterol education program expert panel on detection, evaluation, and treatment of high blood cholesterol in adults. The expert panel. *Arch. Intern. Med.* 1988; **148**: 36–69.
5. Shepherd J, Betteridge DJ, Durrington P, *et al.* Strategies for reducing coronary heart disease and desirable limits for blood lipid concentrations: Guidelines of the British Hyperlipidaemia Association. *Br. Med. J.* (Clin. Res. Ed).1987; **295**: 1245–1246.
6. Strategies for the prevention of coronary heart disease: A policy statement of the European Atherosclerosis Society. *Eur. Heart J.* 1987; **8**: 77–88.

7. Guidelines for the detection of high-risk lipoprotein profiles and the treatment of dyslipoproteinemias. Canadian Lipoprotein Conference Ad Hoc Committee on Guidelines for Dyslipoproteinemias. *CMAJ* 1990; **142**: 1371–1382.

8. Beaglehole R, Jackson R, Stewart A. Diet, serum cholesterol and the prevention of coronary heart disease in New Zealand. *N Z. Med. J.* 1988; **101**: 415–418.

9. Rossouw JE, Steyn K, Berger GM, *et al.* Action limits for serum total cholesterol. A statement for the medical profession by an ad hoc committee of the Heart Foundation of Southern Africa. *S. Afr. Med. J.* 1988; **73**: 693–700.

10. Grundy SM, Stone NJ, Bailey AL, *et al.* 2018 AHA/ACC/ AACVPR/ AAPA/ ABC/ACPM/ADA/AGS/APhA/ASPC/NLA/PCNA guideline on the management of blood cholesterol: A report of the American College of Cardiology/American Heart Association Task Force on Clinical Practice Guidelines. *Circulation* 2019; **139**: e1082–e1143.

11. Packard C, Chapman MJ, Sibartie M, Laufs U, Masana L. Intensive low-density lipoprotein cholesterol lowering in cardiovascular disease prevention: Opportunities and challenges. *Heart* 2021; **107**: 1369–1375.

12. Puri R, Mehta V, Duell PB, *et al.* Proposed low-density lipoprotein cholesterol goals for secondary prevention and familial hypercholesterolemia in India with focus on PCSK9 inhibitor monoclonal antibodies: Expert consensus statement from Lipid Association of India. *J. Clin. Lipidol.* 2020; **14**: e1–e13.

13. Sampson M, Ballout RA, Soffer D, *et al.* A new phenotypic classification system for dyslipidemias based on the standard panel. *Lipids Health Disease* 2021; **20**: 170.

14. Scientific Steering Committee on behalf of the Simon Broome Register Group. Mortality in treated heterozygous familial hypercholesterolaemia: Implications for clinical management. *Atherosclerosis* 1999; **142**: 105–112.

15. Thompson GR. Atherosclerosis in cholesterol-fed rabbits and in homozygous and heterozygous LDL receptor-deficient humans. *Atherosclerosis* 2018; **276**: 148–154.

16. Lambert G. Unravelling the functional significance of PCSK9. *Curr. Opin. Lipidol.* 2007; **18**: 304–309.

17. Seidah NG, Prat A. The biology and therapeutic targeting of the proprotein convertases. *Nat. Rev. Drug Discov.* 2012; **11**: 367–383.

18. Navarese EP, Kolodzieiczak M, Schulze V, *et al.* Effects of proprotein convertase subtilisin/kexin type 9 antibodies in adults with hypercholesterolemia: A systematic review and meta-analysis. *Ann. Intern. Med.* 2015; **163**: 40–51.

19. Raal FJ, Stein EA, Dufour R, *et al*. PCSK9 inhibition with evolocumab (AMG 145) in heterozygous familial hypercholesterolaemia (RUTHER-FORD-2): A randomised, double-blind, placebo-controlled trial. *Lancet* 2015; **385**: 331–340.
20. Koren MJ, Scott R, Kim JB, *et al*. Efficacy, safety, and tolerability of a monoclonal antibody to proprotein convertase subtilisin/kexin type 9 as monotherapy in patients with hypercholesterolaemia (MENDEL): A randomised, double-blind, placebo-controlled, phase 2 study. *Lancet* 2012; **380**: 1995–2006.
21. Blom D, Hala T, Bolognese M, *et al*. A 52-week placebo-controlled trial of evolocumab in hyperlipidemia. *N Engl. J. Med.* 2014; **370**: 1809–1819.
22. Sabatine MS, Giugliano RP, Wiviott SD, *et al*. Efficacy and safety of evolocumab in reducing lipids and cardiovascular events. *N. Engl. J. Med.* 2015; **372**: 1500–1509.
23. Raal FJ, Hovingh GK, Blom DJ, *et al*. Long-term treatment with evolocumab added to conventional drug therapy, with or without apheresis, in 106 homozygous familial hypercholesterolaemia patients: An interim subset analysis of the open-label TAUSSIG study. *Lancet Diabetes Endocrinol.* 2017; **5**: 280–290.
24. Ray KK, Stoekenbroek RM, Kallend D, *et al*. Effect of an siRNA therapeutic targeting PCSK9 on atherogenic lipoproteins. *Circulation* 2018; **138**: 1304–1316.
25. Cuchel M, Meagher EA, du Toit Theron H, *et al*. Efficacy and safety of a microsomal triglyceride transfer protein inhibitor in patients with homozygous familial hypercholesterolaemia: A single-arm, open-label, phase 3 study. *Lancet* 2013; **381**: 40–46.
26. Gaudet G, Gipe DA, Pordy R, *et al*. ANGPTL3 inhibition in homozygous familial hypercholesterolemia. *N Eng. J. Med.* 2017; **377**: 296–297.
27. Myant NB. *The Biology of Cholesterol and Related Steroids*. London: Heinemann; 1981.
28. Bacon F. *The Advancement of Learning*. First published in 1605; revised edition: New York: Modern Library; 2001.
29. Ravnskov U, de Lorgeril M, Diamond DM, *et al*. LDL-C does not cause cardiovascular disease: A comprehensive review of the current literature. *Exp. Rev. Clin. Pharmacol.* 2018; **11**: 959–970.

Chapter 19

The Future of Clinical Lipidology

19.1. A Comparison of Lipid Research on Both Sides of the Atlantic

Although scientists of many nations contributed to the testing of the lipid hypothesis, two nationalities predominated. Almost 50% were American and over 30% were British, with single scientists from other nations comprising the remainder.

Anitschkow published the results of his rabbit experiments in Russia in 1913, but it was not until 1949 that Gofman began his ground-breaking studies on lipoproteins in California. He and his fellow Americans Dawber, Havel and Keys dominated research into the role of cholesterol in atherosclerosis in the 1950s, the lone British contributors during that decade being Oliver and Boyd. In the 1960s, the US momentum was maintained by Kannel in Framingham and by Fredrickson and Levy at the NIH, while the UK was represented by Myant in London. The 1970s were marked by criticisms of the lipid hypothesis by McMichael, Mitchell and Yudkin in the UK, but these were overshadowed by the epoch-making discoveries of the LDL receptor by Goldstein and Brown in Dallas and of compactin by Endo in Tokyo.

The end of the 1980s marked the beginning of the statin era with the licensing of lovastatin in the USA, followed in the 1990s by the statin trials conducted by Pedersen in Scandinavia and by Shepherd and Collins in the UK. This era culminated in the decisive meta-analysis of the statin trials by the international writing committee of the Cholesterol Treatment Trialists (CTT) Collaboration in 2005.

Major scientific awards to scientists whose work features in this book comprised two American recipients of both the Nobel Prize and the Lasker Award; one Japanese recipient of both the Lasker Award and Gairdner Award; and one UK and three US recipients of Gairdner Awards. Thus, 7 of the 10 prizes went to Americans. Overall, they dominated the proving of the lipid hypothesis, both in terms of priority and scientific recognition. Possible reasons for this are considered in what follows.

19.2. The Flexner Report

In 1910, the Carnegie Foundation commissioned Abraham Flexner to ascertain the facts concerning medical education and medical schools in the USA and Canada.[1] His report revealed overproduction of uneducated and ill-trained medical practitioners, mainly due to the existence of numerous poorly-run commercial medical schools. This situation underlined the need for a much smaller number of better equipped and better managed medical schools like the Johns Hopkins Medical School in Baltimore, the first medical school in America run along university lines. The backbone of the proposed institutions was to be the clinical department of internal medicine under the direction of a professor, as was common in German universities, with staff who devoted themselves full-time to teaching and research. Provision of an adequate salary for the faculty would eliminate the need for private practice.

The publication of the report transformed the nature and organisation of medical education in the USA. It assigned medical schools to three categories: the first compared favourably with the Johns Hopkins model; the second was considered sub-standard but salvageable with financial assistance: the third category was of such poor quality that closure was indicated, a fate which befell one-third of all American medical schools in the aftermath of the report.[2]

Flexner's concept of the ideal physician was someone whose animating force was science. He considered that advancement of knowledge by research should take precedence over all other aspects of a physician's life. The Johns Hopkins was his gold standard, where all students spent their first 2 years studying basic laboratory sciences before starting their clinical training. However, the great Canadian physician William Osler, who was one of the founding fathers of the Johns Hopkins before he moved to Oxford, disagreed with Flexner's blinkered approach and considered that the welfare of patients and education of students were just as important as undertaking research, a compromise which most academic physicians adopt nowadays.

Despite its critics, the Flexner report had an enormous and generally positive influence on medical education in North America, especially the need for physicians to have a sound scientific basis for their clinical research. In the UK the department of medicine at the Hammersmith was the only one to emulate the Flexnerian pattern of full-time academic research and teaching, the exception that proves the rule.

19.3. Financial Support for Research

The total expenditure in 2013 of the top three non-commercial funders of health research in the USA (NIH, US Department of Defense and the Howard Hughes Medical Institute) and in the UK (MRC, Wellcome Trust and Department of Health) was $27,851 million and $2,722 million, respectively.[3] After adjusting these sums to account for the fivefold difference in population size, the US expenditure per capita was twice that of the UK. If one restricts the comparison to research into cardiovascular disease, the current expenditure of the NHLBI is $3,845 million per annum. Assuming that two-thirds of this sum is allocated to heart research, again this is twice as great per capita than the $247 million estimated by the UK Health Research Analysis to have been spent on research into cardiovascular disease in the UK in 2018.[4] Hence, one may deduce that the funding of research into atherosclerosis is significantly more generous in the US than in the UK, especially if one takes into account the enormous sums spent by the US pharmaceutical industry on the development and testing of lipid-lowering drugs.

19.4. Training and Accreditation in Clinical Lipidology

The US National Lipid Association (NLA) was founded in 1997 with Virgil Brown as its first president, who was also the founding editor of its journal, the *Journal of Clinical Lipidology*, published by Elsevier. The NLA holds annual scientific sessions and clinical lipid updates, and it organises Master's in Lipidology and Foundations of Lipidology courses in both face to face and online modes. Certification in lipidology in the

USA is available via the American Board of Clinical Lipidology and the Accreditation Council in Clinical Lipidology.

In a similar manner, the European Atherosclerosis Society (EAS) runs a one-year online course that leads to a Certificate of Excellence in Lipidology, accredited by the European Union of Medical Specialists. The Programme is intended for physicians in internal medicine, cardiology, diabetology and endocrinology. It is also relevant to general practitioners and to nurses working in lipid clinics. Like the NLA, the EAS has its own journal, *Atherosclerosis,* which is also published by Elsevier. Hence, both in the USA and Europe there are well-recognised pathways of training, publication and accreditation in lipidology.

The UK also has similar societies that cater to lipidologists, namely HEART UK and the British Atherosclerosis Society, but no affiliated journal. Pursuit of a career in lipidology in the National Health Service is not easy for physicians because Metabolic Medicine has now been incorporated into Chemical Pathology (clinical biochemistry) and it is no longer possible to enter training in this subspecialty under the aegis of the Joint Royal Colleges of Physicians Training Board.[5]

Instead, it entails five and a half years of speciality training in chemical pathology and metabolic medicine, lipids being only one of several topics comprising the metabolic medicine component, the others being diabetes, metabolic bone disease, nutrition, inborn errors of metabolism, thyroid disease and renal stones. There is an urgent need for some form of accreditation in lipidology for physicians in the UK who are not chemical pathologists. The establishment by HEART UK of a postgraduate diploma in clinical lipidology was proposed more than 10 years ago but was never enacted.[6]

19.5. Conclusions

A comparison of lipid-related activities between the US and UK leads to the conclusion that the training of physicians is more scientific, funding for research is more generous and accreditation in lipidology is available in the USA but not in the UK. Collectively, these advantages provide an explanation for the US superiority in lipid research during the latter half of the 20th century.

Marked improvements in training at both the undergraduate and postgraduate levels are needed in the UK if clinical lipidology is ever to achieve recognition as an important medical sub-speciality rather than becoming a minor branch of chemical pathology. The large increase in referrals to lipid clinics brought about by screening for FH and other severe hyperlipidaemias underlines the need for clinical lipidologists possessing both diagnostic and therapeutic skills. The introduction of a suitably approved form of postgraduate accreditation in lipidology for interested physicians with a variety of other backgrounds, such as endocrinology and diabetes, cardiology, and clinical pharmacology could provide the answer to this need.

Future prospects for research into lipids and atherosclerosis in the UK look brighter than they do for clinical lipidology. The UK Biobank, under the leadership of Sir Rory Collins, provides a unique opportunity to conduct Mendelian randomisation studies of the causal basis of genetically determined cardiovascular risk factors. In addition, the Francis Crick Institute and British Heart Foundation have just set up a partnership to support collaborative research into cardiovascular disease, using the unique skills and techniques available at the Crick Institute. The Medical Research Council, Cancer Research UK and the Wellcome Foundation recently announced the award of £1 billion to the Crick, which according to its director Sir Paul Nurse "is an investment that promotes UK science."[7]

19.6. A Cri de Coeur

There is still an important role in Britain for clinicians who can recognise and treat the phenotypic expression of lipid disorders, attributes that were summed up in general terms by the cardiologist Sir Thomas Lewis: "Knowledge that is applied usefully to the health of mankind will almost always come by a series of steps, the first of which is the recognition of the human need, the last of which is the application of a test directly to the human problem. It is therefore in the nature of things, however many steps may intervene, that the first and last shall be clinical; as it is also in the nature of things that almost all .... discoveries .... applicable to treatment of disease have their original source in clinical observations."[8]

Recent advances in DNA- and RNA-targeted therapy mean that distinctions are becoming increasingly blurred between traditional drugs that influence phenotypic expression like statins, that lower LDL cholesterol by inhibiting the enzyme HMG CoA reductase and thereby upregulate

LDL receptors, and gene-targeted therapeutic agents such as the siRNA Inclisiran and the CRISPr base-editing Verve-101. Both of these lower LDL by increasing LDL receptor activity, but by a different mechanism to statins, namely by inhibiting PCSK9 gene expression in the liver and hence reducing PCSK9-dependent catabolism of LDL receptors.[9]

The development of PCSK9-inhibitors resulted from studies of families with low LDL levels due to loss-of-function mutations of PCSK9, vindicating the comments made.by Sir Thomas Lewis 90 years ago about the importance of clinical observation. The impending introduction of other novel forms of lipid-lowering therapy such as the antisense oligonucleotide pelacarsen, that lowers Lp(a) by inhibiting translation of apo(a) mRNA, further illustrates the need for up-to-date postgraduate training in clinical lipidology. Arguably, the knowledge gained from the research that proved the lipid hypothesis should feature in the education of every clinician involved in the management of patients with or at risk of developing atherosclerotic cardiovascular disease.

References

1. Flexner A. Medical education in the United States and Canada. From the Carnegie Foundation for the advancement of teaching, bulletin number four, 1910. *Bull World Health Organ.* 2002; **80**: 594–602.
2. Duffy TP. The Flexner report — 100 years later. *Yale J. Biol. Med.* 2011; **84**: 269–276.
3. https://health-policy-systems.biomedcentral.com/articles/10.1186/s12961-015-0074-z/tables/1.
4. https://hrcsonline.net/reports/analysis-reports/uk-health-research-analysis-2018/.
5. https://www.jrcptb.org.uk/specialties/metabolic-medicine-sub-specialty.
6. Datta BN, McDowell IFW, Rees A. Integrating provision of specialist lipid services with cascade testing for familial hypercholesterolaemia. *Curr. Opin. Lipidol.* 2010; **21**: 366–371.
7. https://www.crick.ac.uk/news/2022-07-11_ps1billion-invested-in-the-future-of-uk-discovery-science.
8. Lewis T. The relation of clinical medicine to physiology from the stand-point of research. *Br. Med. J.* 1932; **2**(3753): 1046–1049.
9. https://media.nature.com/original/magazine-assets/d43747-020-01180-3/d43747-020-01180-3.pdf.

Glossary of Scientific Terms and Abbreviations

^{14}C-cholesterol: Cholesterol labelled with radioactive carbon

^{125}I – LDL: LDL labelled with radioactive beta-emitting ^{125}I

^{131}I – LDL: LDL labelled with radioactive gamma-emitting ^{131}I

α lipoprotein: Electrophoretic terminology for HDL

β lipoprotein: Electrophoretic terminology for LDL

ω-3 fatty acid: Polyunsaturated fatty acid with three or more double bonds

Absolute catabolic rate (ACR): Mass of protein degraded per unit of time

Acetyl LDL: Chemically modified LDL

Acylcholesterol acyltransferase (ACAT): Enzyme that esterifies cholesterol in macrophages

Alirocumab: Monoclonal antibody to PCSK9

Amyloid-β (Aβ): Pathognomonic hallmark in brain of Alzheimer's disease

Analytical ultracentrifuge: Device used to separate lipoproteins according to flotation rate

Androsterone: Androgenic compound

Angiopoietin-like 3 (ANGPTL3): Endogenous inhibitor of lipoprotein lipase

Antisense oligonucleotide: Synthetic nucleotides complementary to mRNA of target gene

Aorta: Large artery conveying oxygenated blood from the heart to the rest of the body

Aortic root: Part of aorta adjacent to heart

Aortic valve: Tricuspid valve situated in aortic root

apoB$_{100}$: Main form of apoB in blood, secreted by the liver, as distinct from intestinal apoB$_{48}$

apoE Leu167del: Rare mutation of apoE gene that causes FH

Arginine: An amino acid

Arteriosclerosis: Synonymous with atherosclerosis

Arteriosclerotic heart disease (ASHD): Synonym of ACVD

Atherogenesis: Pathological process resulting in atherosclerosis

Atherosclerosis: Inflammatory disorder of arteries initiated by cholesterol deposition

Atherosclerotic cardiovascular disease (ACVD): Cardiovascular sequelae of atherosclerosis

Carcinogenic: Cancer causing

cDNA: Complementary DNA

Chlorophenoxyisobutyrate: Active constituent of clofibrate

Cholesterol ester: Cholesterol esterified with a long chain fatty acid

Cholesterol ester transfer protein (CETP): Mediates exchange of cholesterol for triglyceride

Cholestyramine: Anion exchange resin that binds bile acids and lowers cholesterol

Clathrin-coated pit: Mediates internalisation of receptor-bound LDL within cell

Clofibrate: Triglyceride-lowering and HDL-raising agent

Colchicine: Anti-inflammatory drug used to treat acute gout

Colestipol: Bile acid sequestrant, similar to cholestyramine

Compactin: First HMG CoA reductase inhibitor to be discovered

Continuous flow centrifugal cell separator: Device used to effect plasma exchange

Corneal arcus: White-coloured ring around the ocular iris

Coronary angiography: Radiological visualisation of opacified coronary arteries

Coronary artery disease (CAD): Atherosclerosis of coronary artery

Coronary heart disease (CHD): cardiovascular consequences of CAD

Coronary ostia: Origins of left and right coronary arteries in aortic root

Cyclohexanedione: When linked to apoB, it prevents receptor-mediated uptake of LDL

Cytochrome P450: Enzyme system involved in hepatic metabolism of certain drugs

Cytokines: Proteins involved in mediating immune response and inflammation

Diabetic acidosis: Uncontrolled diabetic ketosis leading to coma

Disulphide bridge: Covalently linked sulphide atoms that stabilise conformation of proteins

Double-blind study: Investigators and participants are both blinded throughout the trial

Dyslipoproteinaemia: Generic term for disorders of plasma lipoproteins

Electrocardiogram (ECG): Device used to monitor electrical activity of the heart

Endothelial cell: Cell lining the inner surface (intima) of blood vessels

Epitope: Part of an antigen to which antibodies bind

Essential hyperlipaemia: Original term for severe hypertriglyceridaemia (type 1)

Evinacumab: ANGPTL3-inhibiting monoclonal antibody

Evolocumab: PCSK9-inhibiting monoclonal antibody

Ezetimibe: Drug that lowers cholesterol by reducing its absorption from intestine

Familial defective apoB (FDB): Mutation of apoB that prevents its binding to the LDL receptor

Familial hypercholesterolaemia (FH): Monogenically inherited disorder of LDL metabolism

FDA: Food and Drug Agency

Fibrinolysis: Dissolution of the fibrin component of a blood clot

Fibroblasts: Connective tissue cell often cultured for in vitro studies of lipoprotein metabolism

Foam cell: Cholesterol ester-containing macrophage in atherosclerotic lesion

Fractional catabolic rate (FCR): Proportion of pool of protein degraded per unit of time

Friedewald-estimated LDL: Estimate of LDL cholesterol using the Friedewald equation

Genotype: DNA profile of a specific gene

Glutamine: An amino acid

Glycoprotein: Protein covalently linked to a carbohydrate

Haptoglobin 2-2: Acute phase protein that increases cardiovascular risk in diabetics

Heterozygous FH: Monoallelic inheritance (from one parent) of mutant gene

High-density lipoprotein (HDL): Smallest, most dense lipoprotein (d 1.063–1.21) in plasma

HDL-2: Larger, less dense sub-fraction of HDL (d 1.063–1.125)

HDL-3: Smaller, denser sub-fraction of HDL (d 1.125–1.21)

HMG-CoA reductase: Rate-limiting enzyme of endogenous cholesterol synthesis

Homozygous FH: Bi-allelic inheritance (from both parents) of mutant gene(s)

Hormone-sensitive lipase: Intracellular enzyme that converts triglycerides to free fatty acids

Hydrophilic: Water soluble

Hypercholesterolaemia: Abnormally high blood cholesterol

Hyperlipidaemia: Abnormally high blood cholesterol and/or triglycerides

Hyperlipoproteinaemia: Abnormally high blood lipoproteins (Fredrickson types 1–5)

Hypertriglyceridaemia: Abnormally high blood triglycerides

Hypothyroid: Thyroid hormone deficiency

Ileal bypass: Surgical exclusion from intestinal continuity of bile acid absorbing part of ileum

Inclisiran: Small interfering (si) RNA that inhibits translation of PCSK9 by the liver

Intermediate-density lipoprotein (IDL): VLDL remnant particles (d 1.006–1.019)

Isoforms: Similar proteins produced by a single gene, e.g. apoE 2, 3 and 4

Kringle IV type 2: Number of these in apo(a) are inversely correlated with Lp(a) concentration

Lactone: Carboxylic ester of a compound

LDL receptor: High affinity binding site on cell surface for apoB and E-containing lipoproteins

Lecithin cholesterol acyltransferase (LCAT): Enzyme that esterifies cholesterol in plasma

Left ventricular hypertrophy: Enlargement of left ventricle of the heart

Linoleic acid: A polyunsaturated fatty acid with two double bonds

Lipidology: Study of lipids involved in health and disease

Lipid Research Clinics (LRC): US clinical research centres funded by the NHLBI

Lipoid: Obsolete term for lipid

Lipophilic: Lipid soluble

Lipoprotein (a) (Lp(a)): An LDL molecule covalently linked to apo(a)

Lipoprotein apheresis: Selective removal from plasma of apoB-containing lipoproteins

Lipoprotein lipase: Enzyme that converts triglycerides into free fatty acids in plasma

Lipoprotein phenotypes: Clinical and biochemical characteristics of lipoprotein disorders

Lomitapide: Pharmacological inhibitor of MTP that blocks apoB secretion by the liver

Low-density lipoprotein (LDL): The major transporter of cholesterol in plasma (d 1.019–1.063)

Lovastatin: First HMG CoA reductase inhibitor to be licensed for clinical use

Lymphoblastoid: Lymphocytes immortalised by infection with Epstein Barr virus

Lysophosphatidylcholine: Product of metabolism of phosphatidylcholine by phospholipase A2

Lysosome: Intracellular vesicle containing enzymes that digest proteins

Macrophage: Monocyte-derived phagocytic cells that play a major role in atherogenesis

MRC: Medical Research Council

Metabolic syndrome: Constellation of risk factors for atherosclerosis associated with obesity

Mevalonate: Immediate product of HMG CoA reductase action on HMG CoA

Mevastatin: Synonymous with compactin

Mevinolin: Original name of lovastatin

Microsomal triglyceride transfer protein (MTP): Mediates assembly of chylomicrons and VLDL

Monacolin K: Version of lovastatin discovered by Endo

Monounsaturated fat: Triglyceride containing fatty acids with one double bond

Myocardial infarct: Heart attack, usually due to coronary atherosclerosis

Myopathy: Muscle damage

NIH: National Institutes of Health

Nephrotic syndrome: Renal disorder characterised by proteinuria and hypercholesterolaemia

Niacin: Nicotinic acid

Non-HDL cholesterol: Total cholesterol minus HDL cholesterol

Normocholesterolaemic: Normal blood cholesterol level

Palmar striae: Yellowish pigmentation in palmar creases seen in type III

Paper electrophoresis: Separation of proteins on buffer-soaked filter paper by electric current

Pelacarsen: Antisense oligonucleotide to apo(a) mRNA for lowering Lp(a)

Phase 2: Short trial of drug in target patients to assess efficacy and safety

Phase 3: Longer, larger double-blind comparison of drug vs placebo in target patients

Phosphatidylcholine: Common phospholipid, integral to cell membrane structure and function

Plant sterols: Plant-derived sterols that lower cholesterol by blocking its absorption

Plasma exchange: Removal of plasma by apheresis and replacement with albumin

Plasmapheresis: Separation and removal of plasma from blood (apheresis = take away)

Plasminogen: Inactive precursor of plasmin that lyses fibrin clots after activation

Platelets: Small blood cells that initiate clotting in blood vessels

Pravastatin: Hydrophilic HMG CoA reductase inhibitor

Pre*β* lipoprotein: Electrophoretic terminology for VLDL

Preparative ultracentrifuge: Device used to isolate lipoproteins according to their density

Primary prevention: Prevention of cardiovascular disease in healthy persons

Polyunsaturated fat: Triglyceride consisting of fatty acids with two or more double bonds

Probucol: Lipid-lowering drug with anti-oxidant properties

Proprotein convertase subtilisin kexin (PCSK9): Promotes lysosomal degradation of LDL receptor

***p*-tau:** Phosphorylated proteins that form insoluble tangles in brain in Alzheimer's disease

Receptor-defective: Mutations causing partial deficiency of LDL receptors in homozygous FH

Receptor-negative: Mutations causing complete deficiency of LDL receptors in homozygous FH

Refsum's disease: Autosomal recessive neurological disorder due to phytanic acid excess

Regression studies: Angiographic investigation of effects of lipid-lowering on atherosclerosis

Remnant particles: Particles resulting from action of lipoprotein lipase on VLDL and chylomicrons

Revascularisation: Procedures used to improve blood supply to heart, e.g. angioplasty, bypass

Rhabdomyolysis: Severe muscle damage with release of breakdown products into blood

Saturated fat: Triglyceride consisting of fatty acids with no double bonds

Scavenger receptor: Macrophage receptor that takes up modified LDL in unregulated manner

Secondary prevention: Prevention of CVD events in persons with cardiovascular disease (CVD)

Serotonin: Neurotransmitter that favourably influences mood

S_f 0–12: Flotation rate of LDL in an analytical ultracentrifuge

S_f 12–20: Flotation rate of IDL

S_f 20–400: Flotation rate of VLDL

Simvastatin: Lipophilic HMG CoA reductase inhibitor

Sinuses of Valsalva: Areas bounded by aortic valve leaflets and aortic root wall

Specific activity: Ratio of radioactivity to mass of labelled compound, e.g. dpm/mg

Statin: Generic term for an HMG CoA reductase inhibitor

Svedberg flotation rate (S_f): Negative sedimentation rate in analytical ultracentrifuge

Synvinolin: Obsolete term for simvastatin

Tendon xanthoma: Palpable cholesterol-rich deposits in tendons, especially Achilles

Thyrotoxicosis: Overactive thyroid gland

Thyroxine: Thyroid hormone

***Trans* fatty acid:** Partially hydrogenated (hardened) fat

***t*-test:** Statistical test of significance of difference between means

Tubero-eruptive xanthoma: Yellow, cholesterol-rich protuberances on knees and elbows

Type III hyperlipoproteinaemia: Raised cholesterol and triglycerides due to apoE2 homozygosity

Very-low-density lipoprotein (VLDL): Triglyceride-rich precursor of LDL secreted by the liver

WHO: World Health Organization

Xanthelasma: Yellowish streaks on eyelids

Xanthoma: Cholesterol or triglyceride-rich deposits in skin or tendons indicative of hyperlipidaemia

Xanthoma tendinosum: Latin name for tendon xanthoma

Xanthoma tuberosum: Latin name for tubero-eruptive xanthoma

Index